Power Systems

E. Fumagalli · L. Lo Schiavo · F. Delestre

# Service Quality Regulation in Electricity Distribution and Retail

E. Fumagalli · L. Lo Schiavo · F. Delestre

# Service Quality Regulation in Electricity Distribution and Retail

With 25 Figures and 28 Tables

Springer

Elena Fumagalli
Politecnico di Milano
Milano
Italy

Luca Lo Schiavo
Autoritá per l'Energia Elettrica e il Gas
Milano
Italy

Florence Delestre
Commission de Régulation de l'Énergie
Paris
France

Publication sponsored by:

Council of European Energy Regulators

European University Institute
Robert Schuman Centre for Advanced Studies

ISBN 978-3-642-09251-0 e-ISBN 978-3-540-73444-4



Springer is a part of Springer Science+Business Media
springer.com

Softcover reprint of the hardcover 1st edition 2007



Cover design: deblik, Berlin

# Contents

# Preface

A widespread reform has brought freedom of choice, market opening and privatization to the network services of most countries in a remarkably short period of time, i.e., a couple of decades. A widespread worry that has accompanied this process is related to the effect that a generalized prevalence of the profit motivation in activities inevitably monopolistic or characterized by forms of market dominance could have on the quality of the services provided. Advocates of a direct or indirect state ownership of the utilities companies have long and strongly emphasized the risk that profit-oriented managers and owners would neglect investment not strictly necessary to the creation of revenue.

Liberalization has often (though not always) been accompanied by privatization. Even when a company is not privatized, it will be run as a private enterprise. The era in which large services requiring a highly sophisticated organization are provided by branches of the public administration, or by public entities, is over. The company is now generally accepted as the most appropriate organizational model. This implies that the bottom line of the profit and loss accounts is used as a powerful indicator of managerial efficiency, even in state-owned enterprises. As the state becomes a more careful shareholder and devotes ever closer attention to economic results, the profit motivation emerges even in nationalized companies; and with it, the risk of a deterioration of quality.

In the electricity service, the reform process is well under way, and in many countries it is quite advanced; yet there are no signs of a loss of quality. On the contrary, some countries can show evident improvements. This is a noteworthy development, and regulation has something to do with it.

The process of liberalization has been generally balanced by the creation of new institutions in charge of regulation. The new institutions have faced the issue of the quality of service, as well as other issues, with determination and even some enthusiasm. They have understood that they face a crucial challenge.

Only if the liberalized system can provide better services at lower costs will it be accepted by consumers and voters, and hence only then will it be stable and expected to last, which is a necessary condition to attract investment and thus to provide the conditions for lasting security of supply and further improvements in quality and efficiency. If one of these conditions does not materialize, then the whole process is at risk, and the appeal of the opposite option gains new strength, i.e., the option to go back and put the collective welfare into the hands of one large and vertically integrated company endowed with a *de facto* monopoly power and strictly linked to the government. Competition and regulation can reinforce each other in a virtuous spiral: setting this spiral into motion is the challenge that policy makers and particularly regulators must face in the early stages of reform.

Key to a successful regulation of quality in a liberalized and at least partially privatized system is a large and well-conceived use of incentives. Rather than getting involved in specific decisions on investment and organization, and rather than using rigid command-and-control techniques, the regulator will engineer a mechanism by which good and/or improving quality is remunerated and bad/insufficient quality is penalized. Incentive-based regulation is the new watchword.

A fresh start is essential but not sufficient. Measurement standards must be decided and adopted. An acceptable, equitable end effective set of compulsory measures, incentives and penalties can only be developed on the basis of studies, comparisons and consultation. Regulation is a 'long haul' job. It requires an accumulation of specific knowledge that can only be the fruit of a lengthy and systematic effort in the specific electrical system, with an eye to theoretical developments and to the experience of other systems.

Here is where research, training courses and supplementary study materials, like this handbook, turn out to be necessary.

A major effort is required to spread knowledge acquired on the basis of national experiences to a broader audience, and progressively to promote a shared methodological approach, cooperation among regulators, and a benchmarking of regulatory systems worldwide. The task is huge. The present handbook provides an initial and quite substantial contribution.

This book has not been written in an ivory tower. It is the result of a collective effort by scholars and practitioners working together for many years. The authors have been engaged in the actual making of service quality regulation, in consultation with stakeholders, in systematic comparative research, in academic study and in the training of staff of regulatory institutions and of electricity companies involved in the provision of service.

An interesting aspect is that most of the training in connection with these efforts has been conducted at a European level, with participants from many countries, most of them Member States of the European Union, but with other participants as well. The discipline of quality regulation is new and it is being developed internationally, in a common European frame with increasing relations with researchers and practitioners of other continents. This is quite common in scientific research, but much less so in the training of public administrations. Yet the future improvement of public service depends on an increased use of common efforts and comparative exercises. Only an international benchmarking can grant public administrations some of the stimulus and spreading of knowledge that competition normally yields for companies competing in open markets.

Research, training and the building of a common culture in the field of regulation is the mission of the Florence School of Regulation, which was born as a joint initiative of the Council of European Energy Regulators and of the European University Institute, in the frame of the inter-disciplinary research on European integration which is carried out by the Robert Schuman Centre of Advanced Studies.

We hope that this handbook will be useful to scholars and regulation practitioners in Europe and elsewhere. We are pleased to receive comments and suggestions by users.

*Pippo Ranci*
*Director, Florence School of Regulation*

The Council of European Energy Regulators (CEER) promotes the development of efficient and competitive internal markets for electricity and gas and the setting-up of co-operation, information exchange and assistance among its members. Since its foundation, the CEER has devoted special attention to the quality of service delivered to final customers of electricity networks and has already published three benchmarking reports, in 2001, 2003 and 2005. The CEER benchmarking reports on quality of service reflect the continuously improved surveys and data collected in Europe. The growth in participation by regulators in benchmarking activities (including the new Member States who joined the EU in 2004) has contributed to the consistency and reliability of those reports.

The importance of the regulators' work on quality of service issues, in particular within the context of privatization and liberalization processes, is now widely acknowledged. Service quality regulation is one of the most important issues for regulators: it is imperative to define, measure (and harmonize where necessary) quality indicators, and to monitor and influence performance.

National regulators' contributions for the elaboration of the third benchmarking report showed that many countries are interested in introducing some form of incentive regulation for quality of electricity supply; nevertheless, many did not have sufficient knowledge to embark on such an engagement. Furthermore, the enlargement of the EU introduces new challenges for service quality regulation.

In the light of the CEER's objective of enhancing co-operation, information exchange and assistance among its members, whilst having in mind the countries which have not yet developed significant experience on quality issues, the preparation of a handbook was included in the CEER 2006 work programme in collaboration with the Florence School of Regulation (FSR) in order to promote the expertise needed on this subject and to provide a channel for learning, dissemination and further developments. The publication of this handbook is the proof of a fruitful cooperation between the CEER and the FSR.

The handbook is based on the experiences of countries that have already implemented service quality regulation, and it also takes account of some twinning experiences of the national regulatory authorities of recent accession countries.

The CEER hopes that this work will be a valuable contribution for disseminating the culture of service quality in Europe and around the world, through international institutions interested in regulating services of general interest. Although the theory and practice of incentive regulation for quality of service is presented here for the supply of electricity, the regulation of gas supply or other public services can benefit from this handbook as well.

*Sir John Mogg*
*President, Council of European Energy Regulators*

# Acknowledgments

All members of the CEER Task Force for Quality of Supply, under the direction of Dr. Gabor Szörényi, have contributed to this work with valuable insights and encouragement. We are particularly grateful to those working within the regulatory authorities who provided comments on the manuscript, as well as data, figures and other information.

For the theoretical aspects in this book we owe an enduring debt to our teachers and colleagues. The pedagogical perspective was inspired by the students who attended the classes we taught in different regulatory authorities and at the Politecnico di Milano from 2003 to 2006.

Finally, we are glad to acknowledge the advice and support received from the Florence School of Regulation. In particular, we would like to thank Francesca Davoli for her indispensable assistance, and Clare Tame and Mel Marquis for revising the language.

*Elena Fumagalli*
*Luca Lo Schiavo*
*Florence Delestre*

# 1

# Introduction

## 1.1 Objectives and audience

Service quality is an important issue in the electricity distribution and retail sectors. Customers are highly sensitive to all aspects of service quality and value the speed and accuracy with which their requests are handled, the reliability of the electricity supply, and the characteristics of the supply voltage.

Both theory and empirical evidence indicate that when a regulator chooses to regulate prices using price or revenue caps, a company's incentives to deliver efficient levels of service quality tend to drop. As a consequence, price cap or revenue cap regulation must be supplemented by service quality regulation.

Quality regulation entails complications and subtleties. First of all, there is the multi-dimensional nature of service quality. Furthermore, the ideal level of quality depends on consumer preferences, and these preferences can vary widely among customers. In addition, measuring quality can be difficult, consumer behaviour can affect the quality of the network, and so forth. As a result, there is no simple policy indication for service quality regulation: different means are normally used to induce regulated companies to deliver the desired levels of service quality on different quality dimensions.

The purpose of this handbook is to describe the different regulatory instruments that can be employed in service quality regulation. The focus will be on the design and implementation of quality regulation in practice. A comprehensive theoretical review of the subject is given, for instance, in Baldwin and Cave (1999) and Sappington (2005). Other, more specific theoretical contributions will be indicated in the text.

A second objective of this work is to help develop a shared framework for service quality regulation across the European Union (EU). A higher degree of harmonization in the more conceptual aspects of the regulatory design will facilitate the sharing of ideas and innovations. Similarly, a greater uniformity with respect to some of the more practical issues will encourage the exchange of data and information. In particular, the availability of comparable data could enhance further research and analysis on the subject of applied quality regulation and thus provide sound guidance for regulatory decision-making.

This handbook is intended primarily for junior and senior staff of energy regulatory authorities. In particular it is written for those approaching service quality regulation for the first time. The content of this handbook is also of interest for students attending courses in economics of regulation and public policy. The various aspects of quality regulation treated in this work are suitable for students with backgrounds in economics, engineering, and law as well.

## 1.2 How to use this handbook

The handbook presents the four main regulatory instruments employed today in service quality regulation: *data publication, minimum quality standards, reward and penalty schemes*, and *premium quality contracts.* In the electricity distribution and retail sectors we can identify three main service areas where the regulation applies: *commercial quality, continuity of supply* and *voltage quality.* The structure of the handbook is based on this classification. Accordingly, it describes how the four regulatory instruments can be practically employed in each of the three areas of service.[1] The multi-dimensional nature of quality means that, for each area, we will identify one or, more often, several aspects (or services) that are of interest for customers. In turn, for each aspect (or service) we will define one or more quality dimensions that will be measured and possibly regulated.

Currently service quality regulation tends to focus on regulating *outputs*, namely the measured values (indicators) of those quality dimensions that are most important to customers. However, behind the outputs generated by a regulated company, there is a *process* made up of decisions on investment, network planning and operation, maintenance programmes and asset management. Recent debate suggests that service quality regulation should focus more on the process rather than on the outputs alone. This would imply that the regulator needs to become more involved with asset management and network modelling, in addition to participating in setting network design requirements or emergency procedures. This 'behind the outputs' regulation is one of the possible future developments of service quality regulation. However, for a regulator approaching this topic for the first time, a good understanding of 'output regulation' is an essential step before starting on more innovative paths. For these reasons, this handbook does not include topics such as asset management or network modelling and is organized, rather, according to the three areas of service quality mentioned above.

In particular, Chapter 2 gives an introductory definition of service quality in the electricity distribution and retail sectors and briefly presents the four main regulatory instruments in use today. A general framework for service quality regulation is

[1] The regulation of service quality in relation to the transmission network is also an area of interest for regulators. However, in order to keep the scope of this handbook within certain limits, this subject will not be dealt with in this work although a few references will be given where appropriate.

also presented. Finally, the chapter deals briefly with the main practical tasks that derive from the implementation of the regulatory instruments.

Chapters 3, 4 and 5 deal, respectively, with commercial quality, continuity of supply, and voltage quality. Each chapter follows a similar structure. A first section deals with definitions and the process of data collection in the relevant area. The following sections explain how the main regulatory instruments for quality regulation can be employed for the specific area. The issues presented in those sections are a good representation of the 'state of the art'. As such, they are perfectly adequate and exhaustive for readers taking their first steps in this discipline. More advanced topics are dealt with in a final section of each chapter, in order to give some idea of the most recent developments in the use of the regulatory instruments (normally these developments relate to modifications in the scope or structure of the traditional instruments). These final sections can be skipped by the reader without compromising the understanding of the main topics.

Finally, Chapter 6 contains references on specific issues, focusing in particular on the practical tasks mentioned in the final section of Chapter 2.[2]

## 1.3 General messages

Service quality regulation is still a relatively new matter. However, a number of independent regulatory authorities in EU member states have already accumulated 5-10 years' practice in this area of regulation. This handbook draws a great deal on these experiences, in both practical and conceptual matters. As far as general guidance is concerned, while not exhaustive, the following messages are central for a successful regulation of service quality.

1. Regulatory instruments must be fair and simple to implement. Clear rules on data measurements and collection play a fundamental role in ensuring fairness. Existing, successful regulatory schemes are built upon a relatively small number of quality indicators.
2. Applied regulatory schemes may differ significantly in different countries, as regulation is adjusted to the specific industrial and institutional factors of the country in question. Different policy decisions are all equally valid, as long as the effects of the regulation, in terms of quality improvements versus incurred costs, are satisfactory.
3. Notwithstanding the disparities in design, regulatory schemes share an important common feature: a gradual approach to the implementation process. Such an approach allows regulators to deal with the inherent multi-dimensional nature of quality in an efficient way: beginning with one aspect of quality and

[2] The authors are jointly responsible for the work. However, Chapters 1, 2 and 4 were written by Elena Fumagalli; Chapters 3 and 6 by Elena Fumagalli and Florence Delestre; and Chapter 5 by Elena Fumagalli and Luca Lo Schiavo. Where two authors appear, the work has been equally shared between them. Any conclusions or opinions expressed are those of the authors and not necessarily of their respective organizations.

then enlarging the scope of the regulation only when visible results have been captured and before possible side-effects appear.

4. In this gradual approach, periodic review of the regulatory instruments are vitally important. In practice, quality regulation is never regarded as a permanent solution to all problems. On the contrary, it is subject to periodic evaluation and revision. This allows regulators to learn from practical results and to enlarge and adapt the scope and structure of regulation over time. At the same time, periodic review should give regulated companies the necessary amount of time to make investment decisions under stable regulatory conditions.
5. A regulatory policy for service quality leads to efficient outcomes when the regulator, the regulated companies and customers are able to dialogue in an open manner, with reciprocal trust and acknowledgment of the efforts that each of them is making towards achieving common objectives. In this way, the inevitable mistakes are not only sanctioned and corrected, as indicated by the rules, they also provide an opportunity to learn and introduce improvements. Indeed, quality regulation is characterized by a continuous learning process.

# 2

# The basics of service quality regulation

This chapter outlines the basic elements of service quality regulation. The issues introduced here will be expanded in the rest of the handbook.

Section 2.1 defines the three areas of service quality that are regulated in the sectors of electricity distribution and retail. Section 2.2 describes the four main regulatory instruments used and Section 2.3 summarizes the fundamental elements of these regulatory instruments. A simple model that links the different areas of service quality to the objectives that can be achieved with the different regulatory instruments is presented in Section 2.4. Section 2.5 is an overview of practical regulatory tasks and Section 2.6 gives a few key, practical suggestions.

## 2.1 Areas of service quality

Service quality in the electricity distribution and retail sectors spans a large number of technical and non-technical aspects. The aspects that are normally regulated are conveniently grouped in three areas.

A first area comprises the non-technical aspects that arise in the relationship between the customer and the service providers (mainly the distributor and the retailer), before the beginning of the service and during the validity of the contract. In particular,

- *Commercial quality* covers the quality of a number of services, such as the provision of a new connection (before the supply contract comes into force), as well as meter reading, billing, handling of customer requests and complaints (during the validity of the contract). A distinction between services provided by the distributor (network operator) and those provided by the retailer becomes strictly necessary as a country moves forward to full retail competition. For each of those services a *quality dimension* is identified (normally one per service): for instance, the 'timing' of the new connection or the 'accuracy' in meter reading and billing. Accordingly, the regulatory instruments focus on *indicators* such as the 'waiting time' for the service to be provided, or the 'frequency' of meter readings in a year.

The regulation of these aspects will be dealt with in Chapter 3.

Technical aspects of service quality are generally referred to as 'power quality' issues and include any variations of the voltage or current characteristics around their ideal values. So far, regulation has addressed two main areas of technical quality within the larger field of power quality:

- *Continuity of supply* is related to interruptions of supply. In other words, it focuses on the events during which the voltage at a customer connection drops to zero.[1] Continuity of supply is described by two *quality dimensions*: the number of interruptions and their duration. Accordingly, the regulatory instruments focus on *indicators* of 'frequency' and 'duration' of interruptions.
- *Voltage quality* covers a subset of the possible variations of the voltage characteristics from the desired values. Examples of these variations are voltage dips, voltage harmonics, and flicker. The *quality dimensions* that are relevant for customers in this area are, for instance, the number of such deviations in a period of time or the amplitude of the deviations. Accordingly, regulation focuses on *indicators* such as the 'frequency' of the events, or the distance, in 'percentage', from the nominal values.

These technical areas of service quality mainly concern the network operators (both in transmission and distribution). Regulation of continuity of supply and of voltage quality are dealt with in Chapters 4 and 5 respectively.

In summary, commercial quality covers several services, continuity of supply is concerned only with interruptions (interruptions, however, are of different kinds), and voltage quality includes a number of different aspects of the same problem. For each of the services or aspects in an area, regulators identify the quality dimension (sometimes the quality dimensions) that are important for customers. In order to regulate these quality dimensions, it is necessary to identify a quality indicator that regulators can observe and quantify. It will be in terms of these quality indicators that a company's performance is measured and consequently rewarded or penalized.

## 2.2 Instruments of quality regulation

As explained above, this handbook addresses multiple aspects of service quality. It is important to note that, regardless of the technical or non-technical nature of these aspects, we focus here on dimensions of service quality that a regulator is able to observe, quantify and verify. In other words, the use of the regulatory

[1] According to the European Norm EN 50160, a supply interruption is a condition in which the voltage at the supply terminals is below 1% of the declared voltage. The declared voltage is normally the nominal voltage of the system (i.e., the voltage by which the system is designated or identified), unless a different voltage is applied, by agreement between the supplier and the customer [CENELEC, 1999].

instruments illustrated in this handbook is constrained by the availability of a uniform (across companies) and reliable (verifiable) set of measured quality indicators describing all the relevant dimensions of quality. When such measures are not available the regulator must give precedence to the collection of data, introducing binding instructions for the companies to follow. Because of their importance and specificity, regulatory instructions for data collection will receive in this text (and should receive in practice) as much attention as the regulatory instruments themselves.

There are basically four instruments that a regulator might employ to secure desirable levels of performance on the multiple dimensions of service quality.

One simple instrument is the *publication of data* on company performance. The dissemination of information regarding a company's performance with respect to a set of quality indicators provides unambiguous incentives for increased investments in quality [Weisman, 2005].

A second relatively simple instrument is the setting of *minimum quality standards* (MQS). The regulator specifies *performance standards*[2] in the form of a minimum level of service quality that a company is expected deliver to its customers. Non-compliance with these minimum levels of quality might entail the payment of penalties. This instrument can be valuable especially when the company tends to deliver particularly low levels of service quality to a group of customers. The appropriate design of MQS requires some knowledge of the costs of supplying quality and the benefits that customers derive from it.

The third instrument is the introduction of a *reward and penalty scheme*. The regulator specifies *performance standards* and implements rewards and penalties for exceeding or failing to achieve these standards. This instrument assumes that the regulator is well informed about the value that customers place on service quality and allows the regulated company some degree of discretion: given the *performance standards* and the associated financial incentives, the company is expected to employ its superior knowledge of costs to deliver an efficient level of service quality [Sappington, 2005]. Reward and penalty schemes are complex to design and to implement. Nonetheless, where already in use, they have delivered extremely positive results on the regulated quality indicators.

The fourth instrument is the promotion of *premium quality contracts* between a company and its customers. Such contracts are especially useful for customers with a high valuation of service quality. They specify individual *performance standards* and prices for the service, as well as individual compensation payments. These are paid by the company if it fails to deliver the quality specified in the contract. In theory, premium quality contracts have the considerable advantage of being able to reveal consumer preferences and to provide incentives for the company to deliver differentiated levels of service quality accordingly. In practice, such contracts have not been widely employed.

---

[2] Performance standards specify the level of quality that the company is expected to supply (see also Section 2.3).

These four instruments can be employed alone, or in combination with one another, in all three areas of service quality described in Section 2.1. In practice, certain instruments have been employed more to some aspects or services than to others.

## 2.3 Basic elements of the regulatory instruments

Before going into the details of how service quality is regulated in practice, it is useful to mention briefly the basic, common elements of the policy instruments described above.

First of all, the instruments of quality regulation require reliable *measures* of the corresponding dimensions of service quality. What is being measured is obviously a quality indicator. The measured quality indicators describe the actual performance of the regulated company. This information is in itself sufficient to make the simplest instrument - data publication - effective.

*Performance standards* are the second element of the other regulatory instruments and specify the level of quality that the company is expected to supply. They represent respectively: a lower limit of service quality delivered to the individual consumer in case of MQS; an average level of quality in the case of reward and penalty schemes; and an agreed-upon individual level of quality in the case of premium quality contracts. Performance standards are given, again, on quality indicators.

A regulated company performing below the performance standard will (in general) be penalized and a company performing above that standard will (sometimes) be rewarded. Hence, *financial incentives* are the third basic element of a regulatory instrument. In the case of MQS and premium quality contracts, financial penalties take the form of monetary compensation paid to the affected customers. By contrast, a functional relation between quality and price is defined in the structure of reward and penalty schemes. In general, the greater the distance of the measured quality indicator from the performance standard, the greater the financial incentive.

As will become clear in the following chapters, the regulatory instruments employed in practice can assume a variety of different forms, but they are always built on these same three basic elements.

## 2.4 A framework for service quality regulation

Service quality regulation can be usefully and synthetically presented using a simple model that links the different areas of service quality to the objectives that can be achieved with the different regulatory instruments.

Acquiring reliable information on the actual level of quality supplied by regulated companies and disseminating this information is the first objective pursued by regulators (*collecting information and making it available*). Drawing the attention of regulated companies as well as consumers to service quality issues, and thus creating incentives for higher company expenditures on quality, can be achieved by means of publication of data on actual company performance and by using cross-company comparisons. Note that, behind publication, there is always a process of data collection according to regulatory instructions and guidance. This ensures the necessary accuracy and reliability of the published information. More importantly, this information is crucial when making decisions on the level of performance standards.

The level of quality provided to individual consumers by a distribution company is generally distributed over a range of values. Regulators are concerned with the possibility that a group of consumers might receive unacceptably low levels of service quality (those on the low end tail of the distribution). Thus, a second objective pursued by regulators is to *protect the worst-served customers*. The instrument of minimum quality standards is well design to respond to this specific goal.

Along with the tail of the distribution of quality levels, regulators are concerned with the average level of quality supplied by the regulated company. Reward and penalty schemes respond well to the objective of inducing the regulated company to deliver, on average, an *efficient level of service quality*. As will be better explained, efficiency implies that the company's decision on the level of service quality accounts for both the company expenditures for higher quality, and the costs borne by the customers for poor quality.

Finally, regulators are concerned with the needs of customers with a high valuation of quality (i.e., higher than average). The objective of regulators in this case is to create *market-like mechanisms* that will give these customers the opportunity to individually negotiate with the regulated company to secure higher service levels. Premium quality contracts enable customers to receive customized levels of service quality (in general, higher than average), with limited involvement on the part of the regulator. Regulators may also favour the *introduction of competition* in service areas where, traditionally, only the regulated companies operated. Examples of these areas are the 'metering' service and the 'connection to the newtwork' service.

A general framework for service quality is thus represented in Table 2.1, which has as many rows as the areas of service quality described in Section 2.1 and as many columns as the above-mentioned objectives. Conceptually, all regulatory instruments described in Section 2.2 can be employed on all dimensions of service quality and the elements in Table 2.1 provide specific examples of such use (the particular regulatory instruments indicated will be described in the corresponding chapters). However, the regulation of each area of quality has followed a different path.[3]

Regulatory instructions on data collection and publication of data, together with MQS, have been employed in commercial quality regulation for quite some time (MQS were introduced for the first time in the United Kingdom (UK), with the 'Citizen's Charter' initiative, in 1992). Reward and penalty schemes have been employed in this area only recently and for very specific aspects of quality (in particular, the quality of call centres); their use is, to date, not widespread. As indicated in Table 2.1, premium quality contracts are not employed in this area. In

[3] An empty space in Table 2.1 signals that a regulatory instrument corresponding to the relevant objective and area has not yet been employed in Europe.

**Table 2.1.** A general framework for service quality regulation

| | Collecting information and making it available | Protecting the worst-served customers | Delivering an efficient level of quality | Favouring market-like mechanisms and competition |
|---|---|---|---|---|
| *Commercial quality* | Regulatory instructions for recording customer requests | Guaranteed Standards on making and keeping appointments | Reward and penalty schemes on the quality of call centres | Competition in providing connections |
| *Continuity of supply* | Regulatory instructions for recording interruptions | Guaranteed Standards on the duration of an unplanned interruption | Reward and penalty schemes on the average interruption duration | Premium quality contracts on the number of unplanned interruptions |
| *Voltage quality* | Measuring campaigns | Voltage quality standards | | Premium quality contracts on supply voltage variations |

turn, although it is not very common yet, a few commercial services can be offered in a competitive environment.

Collection and dissemination of information regarding continuity of supply is a well established practice in Europe. In turn, reward and penalty schemes appeared only after the year 2000 when regulators became concerned with the (possibly perverse) incentives towards cost reduction implied by price cap regulation and privatization. Reward and penalty schemes were introduced in Italy in 2000, in Norway and Ireland in 2001, in the UK in 2002, in Hungary and Portugal in 2003, in Sweden in 2004, and more recently in The Netherlands and Estonia. The advantages provided by this regulatory instrument have created a strong interest on the part of numerous other regulators in Europe [CEER, 2005]. In general, reward and penalty schemes have been employed before MQS: it is easier to measure average continuity indicators than it is to measure continuity for individual customers. At the present time, all regulatory instruments, including MQS and premium quality contracts, are in use.

Voltage quality is an extremely technical and complex subject and, in general, the involvement of regulators in this area is indeed extremely recent. This explains why regulation of voltage quality is still at a preliminary stage compared to the other two areas of service quality. As of today, regulators are predominantly engaged in collecting data. There are only very few cases where regulators have introduced forms of protection for the worst-served consumers (voltage quality standards) and/or regulatory instruments (premium quality contracts) that act on the technical characteristics of the voltage.

In summary, the complexities that arise in regulating quality suggest a gradual approach. It is perfectly acceptable to introduce a regulatory policy that addresses only one or a few of the relevant aspects of service quality, and/or that initially employs only the simpler instruments of quality regulation. A subsequent modification of the regulatory mechanism and of the number of quality dimensions that are regulated can be adopted at a later stage. Indeed, this distinctive aspect of quality regulation - i.e., the scope for gradual measures - is extremely important to keep in mind.

A last observation regards the general validity of the above model. Quality is multi-dimensional in other network industries as well, including the gas distribution sector, the public transportation sector, and the highways sector. In the same way, regulators can always be expected to pursue numerous objectives. Although different aspects of quality will have a significant impact on customer satisfaction in different sectors, the model incorporated in Table 2.1 can provide guidance for the design of efficient and workable policies in several network industries.

## 2.5 Regulatory tasks

The introduction of quality regulation is a complex process that generally expands over a period of years. Indeed, a workable regulatory framework is the result of a series of activities.

First, the regulatory instruments require measures of the relevant dimensions of service quality. The introduction of *instructions and guidance on data collection* is therefore a necessary precondition for any regulatory intervention. Instructions and guidance are useful to ensure that measuring, registration, and reporting requirements are consistent with the purposes of the regulatory framework. These instructions enable the regulator to control the process of data collection, and in the last instance, they promote credibility and fairness with respect to the regulatory framework. In order to propose viable measuring protocols, the regulator has to acquire a fairly good knowledge of several organizational and technical aspects of the electricity distribution (and retail) business. It should be noted that regulated companies may need a couple of years to adopt and adjust to the requirements. In addition, the data provided by network companies should be subject to auditing procedures. Audits can be carried out by regulators, consultants, or even by the companies themselves, according to rules set by the regulator. In any case, there is a broad consensus as to the importance of audits, especially in the first years. The frequency of the audits can then be relaxed over time.

Practical guidance for enforcing measuring protocols for the different areas of service quality will be given in the corresponding chapters. Auditing procedures are dealt with in Chapter 6.

Second, in order to design a viable regulatory instrument the regulator must interact with all interested parties. There are two preferable means of communication: the *consultation process* and *customer surveys*.

It is extremely important for a public institution to make its decisions in an open and fair manner. In this regard, a consultation process has numerous advantages: it improves the quality of the decisions; it facilitates an understanding of the regulator's work; it helps to make the decision-making process clear and transparent to those who are affected by it; and it gives an idea of how regulatory decisions will develop in the future [Ofgem, 2002b].

Several regulators use customer surveys to estimate a number of factors concerning quality of service: customer satisfaction, customer priorities and expectations, and customer willingness to pay (WTP) for quality. Surveying these issues supports the decision-making process in identifying the dimensions of quality that are important for customers, and it helps to design the regulatory instruments. For instance, surveys regarding WTP for improvements in continuity of supply are crucial for defining reward and penalty schemes. In addition, surveys are important for evaluating the results obtained with the regulation.

The consultation process and the use of customer surveys are discussed in Chapter 6.

Third, *assessing* whether *the regulatory instrument* actually delivers the desired results is another essential aspect of the regulatory process. Regulatory schemes should be assessed with respect to at least two variables: their effects on regulated (and non-regulated) quality dimensions and their costs and benefits for companies and customers. In addition, the complexities that arise in regulating quality suggest a gradual approach. Indeed, the design of the regulatory framework (e.g., the number and type of instruments employed, and/or areas of quality to be addressed) as well as of the regulatory instruments (level of the performance standard and/or of the financial incentives) always change over time, as results are assessed and new customer needs emerge. This distinctive aspect of quality regulation will be highlighted throughout the handbook whenever appropriate.

It is important to note that the order followed in the above description of regulatory activities is not accidental. Measurement protocols are a prerequisite for the implementation of each of the regulatory instruments. Customer surveys and the consultation process provide essential contributions to the design of the regulation (definition of the regulatory instruments and choice of the regulated dimensions). Assessments clearly follow the implementation stage. Regulators agree on the benefits of dealing with these tasks in a gradual manner: the expertise they require is rather sector specific and sometimes extremely technical. For this reason (and also because human resources are finite) it is customary for regulatory authorities to collaborate with experts from universities and consulting firms.

## 2.6 Suggestions

One key suggestion for a regulator approaching quality regulation for the first time is to explore the regulatory (and legal) background as well as relevant past experience. This investigation may take several different 'directions'.

A first direction to explore is the *existing regulatory and legal framework*. Customer service standards and network security codes may already exist in the distribution (and/or transmission) licensing contract or elsewhere. Other relevant information regarding existing service quality obligations should be sought in legal codes. In addition, the responsibility for quality regulation is sometimes shared by different regulatory agencies: ministries, governmental agencies, competition authorities, energy inspectorates and so forth. Before designing an instrument of quality regulation, the legal competences and priorities of each agency, including those of the energy regulator, must be taken into account. For instance, an energy regulator may have the power to enforce a regulatory policy, but not to design and ratify it. Or, as another example, the regulator may have the power to set the tariff, but not to impose monetary penalties on the regulated companies. Nonetheless, the regulator can always actively contribute to designing policy and then enforce it in collaboration with the other competent bodies.

Secondly, experience with service quality regulation could be available in *other regulated sectors* other than electricity, e.g., in telecommunications or gas. Understanding the difficulties encountered as well as the successful outcomes achieved in these other areas may facilitate the introduction of quality regulation in the electricity sector.

Finally, *international experience* with quality regulation in the electricity sector is easily accessible. The Council of European Energy Regulators (CEER) issues a periodic *Benchmarking Report on Quality of Electricity Supply.* These reports describe the regulatory instruments adopted in EU member states and compare service quality data.[4] The benchmarking report published in 2005, among other things, compares, for the first time, the main aspects of reward and penalty schemes adopted in EU member states [CEER, 2005]. Detailed information on international practice can be obtained from national regulatory authorities, consulting their websites and reading the texts of the relevant decisions. Further information is obtainable from the academic literature: several regulators have contributed articles in refereed journals and at technical conferences.

4 CEER (www.ceer-eu.org) brings together experiences from member states of the European Union and the European Economic Area.

# 3

# Commercial quality

Commercial quality regulation addresses, for a number of services, the non-technical aspects of service quality that arise in the relationship between customers and service providers.

In practice, commercial quality regulation employs three of the four regulatory instruments presented in Chapter 2: data publication, minimum quality standards and (rarely) reward and penalty schemes. Although there are no conceptual obstacles, in practice premium quality contracts have never been used for commercial quality. For this reason they are not dealt with here.

Section 3.1 defines and explains how to measure commercial quality performance. Section 3.2 deals with publication of data. Section 3.3 describes minimum quality standards. Section 3.4 describes two examples of a reward and penalty scheme applied, in particular, to the quality of call centres. In practice, only a few schemes of this type have been implemented in Europe. More advanced topics are discussed in Section 3.5.

It is important to note that each regulatory instrument is effective in itself. The publication of reliable data on company performance is a simple, yet powerful instrument. Minimum Quality Standards (MQS), together with data publication, are the preferred and most widely employed instruments for commercial quality regulation in Europe. For this reason, it is advisable, especially at the beginning, to focus on designing and implementing MQS for those services that are most important to customers, instead of dealing with the complexities of a reward and penalty scheme. Even these relatively simple regulatory instruments require time to be designed, and they demand continuous work on the part of the regulator and companies. Moreover, they must periodically be revised and adjusted. Experience suggests that it will take at least one to two years' work to introduce commercial quality regulation.

## 3.1 Definitions and data collection

In addressing the non-technical aspects of service quality we need first to identify the services that are most often regulated, their relevant quality dimensions and the measured quality indicators.

As far as services are concerned, it is useful to distinguish between those services that are provided *before the supply* of electricity begins and those that are provided *during the validity of the contract.* Among the latter, it helps to make a distinction between *regular* and *occasional* transactions. Table 3.1 lists a number of such services, according to the above classification.[1] Further details on the most frequently regulated services are given in Section 3.3.

Transactions before supply begins include requests for connection to the network and meter installation, and requests to estimate charges for connections to the grid or for carrying out any related works. The quality dimension of such transactions mainly consists of the time taken to deal with them. The quality indicator that a regulator measures is, thus, the 'waiting time' for the service to be provided (e.g. the number of days taken to connect a new customer).

Regular services provided during the validity of the contract cover billing, meter readings, and services offered by customer centres or call centres. The quality of these transactions mainly consists of their regularity and accuracy. Therefore, the measured quality indicators will be the number of incorrect bills, the frequency of meter readings, customer satisfaction with respect to the precision of the information given at customer centres, and so forth.

Occasional services are often those provided at the customer's request. Customers can request a check for technical failures or disturbances (e.g., with the

**Table 3.1.** Frequently regulated services [CEER, 2003]

| Before supply | During contract validity | |
|---|---|---|
| | Regular | Occasional |
| Providing (supply and meter) | Accuracy of estimated bills | Responding to failures of a distributor's fuse |
| Estimating charges for connection* | Actual meter readings | Voltage complaints |
| Execution of connection-related works* | Service at customer centres | Meter problems |
| | Service at call centres | Appointement scheduling |
| | | Responding to customer requests for information |
| | | Responding to customer complaints |
| | | Reconnection following lack of payment |
| | | Notice of supply interruption |

*Requested also during contract validity

[1] Note that this is only a list of services that are often regulated in EU member states; it is not an indication of services that should be regulated.

meter or the supplied voltage), solicit information, present a request and so forth. As for transactions before supply begins, quality mainly consists of the waiting time for the service to be provided. Thus, the measured indicators generally refer to the time needed for the company to respond, but they may also relate, for instance, to the ability of the company to keep a scheduled appointment.

As far as commercial quality is concerned, it is not only important to define the services that are regulated and the measured quality indicators (the 'waiting time' or 'accuracy' with which the service is provided); it is also necessary to identify the *company* that is subject to the regulatory policy.

Traditionally the distribution network operator provided the transportation and connection services, metering service and electricity retail service. Currently, depending on the level of unbundling prescribed by law, such services could be provided by three or four different companies. At the present time the traditional monopoly distributor is the only provider who can offer transportation. On the contrary, even if it is still quite rare, several companies could offer the connection service (the distributor and other engineering companies). In addition, the metering service can be provided by the distributor or by the retailer, but also by other companies specialized in this line of work. Finally, retailing is becoming an open market activity and several companies, in addition to the local distributor, may be competing to serve customers. This situation implies two main consequences.

First, it is important for a regulator to specify the company to whom the commercial quality regulation applies (including the indications on data collection). Some requirements may apply to all companies (e.g. the speed of responding to customer claims and requests). Some other requirements may only apply to companies operating in a specific market segment (e.g. the requirement concerning the number of annual meter readings will obviously apply only to those companies offering the metering service).

Secondly, depending on the level of liberalization, with the exception of transportation, the above mentioned services may be offered by companies competing with one another. As regards the opportunity to regulate the quality of transactions between a company operating in a liberalized market segment and its customers, competition should induce companies to perform at, or above, a certain minimum level of quality. Because of the novelty of the subject, the decision of whether to regulate the quality of such transactions is mostly left to the regulator's judgement. However, as a general indication, regulation should always apply in the case of services offered by a monopoly company. Similarly, where there are both eligible and captive customers, regulation should also apply to all transactions involving the latter (on this subject see also Section 3.5).

As far as guidance on data collection is concerned, regulatory instructions should be designed to gather reliable information on the performance of regulated companies. The procedures for measuring performance and reporting data to the regulator should be defined clearly and in the same way for all companies, and they should be consistent with the regulatory instruments in use. All measured indicators can be published and some of them will be subject to regulation, most likely in the form of minimum quality standards (MQS).

In most cases, the indicator that describes the non-technical quality of a service is the time between the customer request and the actual provision of the service (the 'waiting time'). Thus, a regulator will normally introduce an obligation for companies to register at least the following for all requests received:

- the type of service requested;
- the category of end-user requesting the service (if the regulator plans to apply MQS with compensation payments that are differentiated per customer type);
- the customer's identification data;
- the date of receipt of the request for the service;
- the date on which the service is carried out.

The most common indicator for commercial quality - waiting time - is derived from the last two items in the register. The register can take different forms, but it is basically a chronological list of services requested by the customers and carried out by the company during a specified period of time (generally a year).

When the service is regulated, a performance standard will be set by the regulator (see Section 2.3). However, the regulation will usually define a set of exemptions from the obligation to pay a financial penalty when the performance standard is not met. A company is not normally responsible for failing to meet the performance standard in situations of *force majeure* (i.e., in exceptional circumstances beyond the company's control, such as severe weather conditions, public authority interventions, calamities, etc.), or when the company can prove third-party responsibility. Setting rules for assessing a company's responsibility is a very sensitive issue: definitions of *force majeure* and third-party responsibility should be as clear as possible in order to avoid litigation.

The information regarding exemptions should be recorded in the register. Thus, for each of the requested services that are subject to regulation the company should indicate the following:

- the cause, among those set out in the regulation (including those leading to an exemption), of any failure to respect the performance standard stipulated for the service requested; or
- the reason and date of any failure to respect the requirement to follow up the request for the service, if this was the case.

In addition, one must bear in mind that the service provider's work may require other types of intervention carried out under the responsibility of the customer or other parties. For instance, it may be necessary to obtain authorizations to carry out any works needed to connect a new customer. The time needed for obtaining authorizations and/or for the customers to complete their share of works should be recorded in the register. For this reason, the company should keep track of:

- dates of submission of requests for authorizations and the dates on which these are obtained;
- date of notification of the completion of works for which the customer is responsible.

These time periods will not be counted as time taken to complete the requested service.

The quality indicator of certain commercial services (mainly regular transactions) is not the waiting time but rather the accuracy of the service or, more generically, customer satisfaction. A typical case of accuracy regards the number of meter readings in a year: the measured indicator is the number of readings. When this service is regulated the company must register the readings made in an appropriate form. Customer satisfaction is an indicator of quality that is mostly used to regulate the performance of call centres or customer centres. Quality indicators for customer satisfaction are often scaled indications of satisfaction with respect to the politeness of staff, their willingness to help, the accuracy and usefulness of the information given, and so on.[2] Data collection on customer satisfaction will be dealt with separately in Section 3.4.

Finally, it is important to ensure that effective arrangements are in place to report the information contained in the register. To this end, the regulator defines a reporting year (e.g. 1 January - 31 December), provides companies with standard templates to be used when reporting information (normally in electronic form), and indicates the deadline for the submission of data. The information provided may be published, and performance across companies can be compared. Therefore, companies should also be informed about the form and means of publication of the reported data. If possible, the regulator should conduct audits on the information registered and reported by companies. Audits should focus on the correctness of the measures and on the compliance with the rules for registration and reporting (see Chapter 6). If audits are carried out, the regulator should indicate when (within a given time-frame) the companies concerned can expect the appointed auditors to undertake an audit.

## 3.2 Publication

Publication of data and comparison of performance across companies are effective ways to stimulate companies' incentives to make quality improvements.

This regulatory instrument can be employed as soon as data (recorded pursuant to the regulator's instructions) are available, even if no other regulatory instrument is in place. When MQS and/or reward and penalty schemes are enforced, the information communicated to the public should include company performance with respect to the performance standards set by the regulator. If a breach of standards results in financial penalties, this information can also be communicated to the public.

In practice there are several channels of communication with the public. First of all, a summary of the information reported by companies can be published in the annual report of the regulatory authority. Published information can be in a statistical form or it may refer to individual companies. Table 3.2 gives an example

[2] Responses are generally given on a scale from 'very dissatisfied' to 'very satisfied'.

**Table 3.2.** Example: commercial quality performance, Italy [AEEG, 2004]

| Service | Number of requests | Standard of performance | Average performance | % of requests above perf. std. |
|---|---|---|---|---|
| Estimating charges for works | 504422 | 20 working days | 11.11 working days | 2.43% |
| Execution of simple works | 437457 | 15 working days | 8.70 working days | 2.31% |
| Connection (supply) | 1664078 | 5 working days | 1.83 working days | 0.91% |
| Deactivation of supply | 767877 | 5 working days | 2.09 working days | 0.51% |
| Reconnection following lack of payment | 310540 | 1 working day (Sat. included) | 0.72 working days | 2.99% |
| Appointment scheduling | 69952 | 3 hours | - | 1.24% |

of the former: the data, which correspond to the year 2004, concern a customer base of 34 million low voltage customers in Italy, and they indicate performance with respect to MQS).

A second means of providing information to the public is to prepare periodic reports on service quality and to make them available for downloading on the regulatory authority's website. Another option is to require companies to report their performance directly to their customers, attaching a note to the bill, once a year. This is efficient because it notifies customers about the level of service quality that they should be receiving. Finally, the regulatory authority's website can be used to host a public-access database with data on service quality.

## 3.3 Minimum quality standards

Minimum quality standards are a relatively simple instrument to secure a minimum level of quality for all customers. In commercial quality regulation they take the form of Guaranteed Standards (GS) and Overall Standards (OS).

*Guaranteed Standards* (GS) refer to service levels (performance standards) which are stipulated by the regulator and which must be met in each individual case. If the company fails to provide the level of service required, it must compensate the customer affected, subject to certain exemptions. A GS relates to a specific service (e.g., connecting a new customer) and it is composed of two elements: the performance standard, which is set at a level of service that is reasonable to expect companies to deliver in all cases (e.g., 10 working days); and a rate of compensation a customer is entitled to receive if the performance standard is not met (e.g., 30 euros).

*Overall Standards* (OS) cover areas of service where the regulator does not consider it appropriate to give individual guarantees, but where customers in general can expect companies to deliver predetermined, minimum levels of service (subject to certain exemptions). An OS refers to a specific service (using an example similar to the one given above, i.e., connecting a new customer when the connection calls for complex works) and establishes the minimum percentage of transactions (e.g., 90%) that must be carried out within a certain time limit (e.g., 40 working days). In contrast to the case of GS, no compensation is paid to customers in cases of breach of the performance standard.

In order to understand the difference between these two regulatory instruments let us assume that the same number of requests (100) have been received for simple connections subject to a GS and for complex connections subject to an OS. Suppose further that, in both cases, 13 connections were realized only after a number of days in excess of the performance standard (respectively, 10 and 40 working days). The time taken to meet the other requests varied but in each case conformed with the performance standard (see Figure 3.1). In actual fact, simple connections were provided, on average, within 7.7 working days, and complex connections were provided within 30.8 working days. Suppose that a total of 6 exemptions apply to the 13 cases exceeding the performance standards, due to *force majeure* (FM) or third-party responsibility (e.g., where users were at fault). In those circumstances, as for the GS, the company would pay 7 compensation payments to affected customers. By contrast, overall company performance with respect to complex connections, satisfies the regulatory requirement: 93% of the customers were in fact connected within 40 working days ($\frac{13-6}{100-6} \cdot 100 = 7\%$).

### 3.3.1 Practical issues

The design of GS and OS requires the regulator to make a number of decisions regarding: regulated services, performance standards, compensation payments and exemptions.

Of all the commercial transactions occurring between electricity customers and distributor (and/or retailers) only a small number are normally subject to regulation. The regulator should make a selection and focus on those *services* that are the most important for customers. A better understanding of customer priorities and expectations can be gained by carrying out a customer survey (see Chapter 6). In the same way, there is no fixed rule for deciding whether to apply a GS or an OS. However, a GS normally concerns transactions that are more frequent, occur in a more standardized form, and are thus more numerous. Secondly, as GS refer to the level of quality of each transaction, imply stricter control with respect to company performance. Consequently, in general, regulators tend to set a small number of both GS and OS. As company performance improves over time, the number of OS gradually drop and the number of GS increase.[3] Alternatively, a regulator can

[3] In fact, depending on circumstances, both GS and OS can be dispensed with altogether, for example, when the service they refer to is offered by companies in a competitive environment.

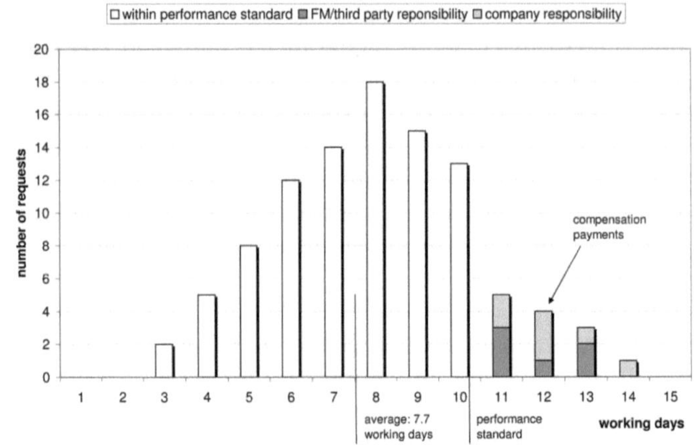


(a) Guaranteed Standards

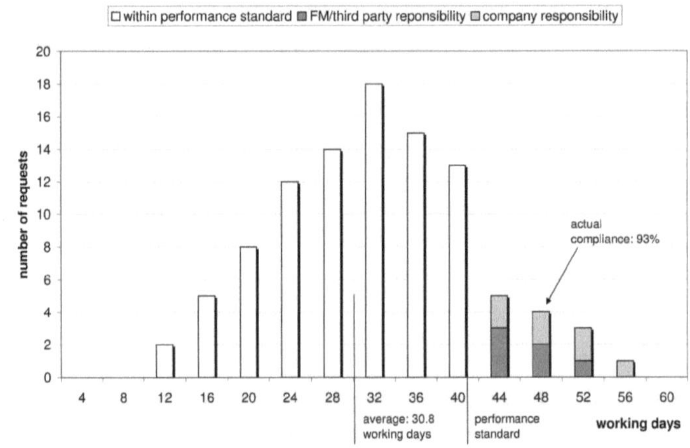


(b) Overall Standards

**Fig. 3.1.** Guaranteed and Overall Standards - an example

rely on the experience of other countries. Examples of GS and OS for commercial quality regulation are given at the end of this section.

*Performance standards* for regulated services are chosen by looking at company performance over time (when available), by consultation with companies and customers, or by relying on international experience. In addition, performance standards should periodically be revised. Regulators should compare the actual, measured performance with the requirements indicated in the regulation. In light of that comparison, a stricter performance standard could be chosen, an OS could become a GS, or in the case of OS, the regulator could require full compliance with the performance standard in 100% of the cases.

As far as GS are concerned, a few decisions regarding *compensation payments* are particularly important. First, such payments can be of two types: automatic or at the customer's request ('on request').

Automatic compensation payments do not depend on customer intervention. If a GS is breached, the distributor will automatically pay compensation to the affected customer. In order to do this, the company must be able to identify the individual customer that has received below-standard service. This should be straightforward if the register is up to date and accurate.

Under an on request compensation scheme, customers must contact the distributor or retailer in order to receive the compensation. If possible, regulators should avoid this approach because experience shows that it weakens the power of the regulatory instrument: not all affected customers request compensation, and distributors may try to discourage customer actions with exaggerated response delays or burdensome legal procedures. On request payments should only be implemented when direct monitoring of individual transactions is not possible. In this case, the regulator can also introduce other mechanisms to encourage customers to request compensation, such as standard forms and clear procedures.[4]

Secondly, regulators decide on the amount of the compensation and the method of payment. The compensation is usually a fixed sum, generally differentiated according to customer category (domestic/non-domestic) and/or voltage level of the connection (low/medium voltage). For example, a compensation scheme could pay 30 euros to a low voltage (LV) domestic customer, 60 euros to a LV non-domestic customer, and 120 euros to a medium voltage consumer. The method of payment should be simple (for instance, a credit in the next bill, or a direct payment). Regulators can introduce a mechanism to protect customers against the risk that the company will not pay timely compensation. For instance, a regulator can fix a maximum time for payments owed under the GS. A company that fails to pay within the fixed maximum time must provided a higher compensation. Alternatively, the regulator can introduce a compensation payment scheme that increases with time. According to the example given in Figure 3.2, if the automatic compensation (AC) is not paid within 90 working days from the performance standard (PS), the company is liable to pay a double compensation. If the compensation is not paid within 180 working days of the performance standard, the company pays 5 times the initial amount.

The *exemptions* from payments can be numerous and complex. As noted above in Section 3.1, *force majeure* refers to exceptional circumstances beyond the control of the distributor/retailer (e.g., severe weather conditions, public authority interventions, calamities and so forth). The definition of *force majeure* is a sensitive issue, and it may differ across countries. Third-party actions can be another cause of exemption from payment, if they can be proved (e.g., where the time needed to obtain an authorization depends on a third party). In commercial quality regulation, the customers themselves can also prevent the service provider from meeting

[4] Hungarian distribution companies can choose between making payments automatically or on request. However, on request payments are larger than automatic ones.

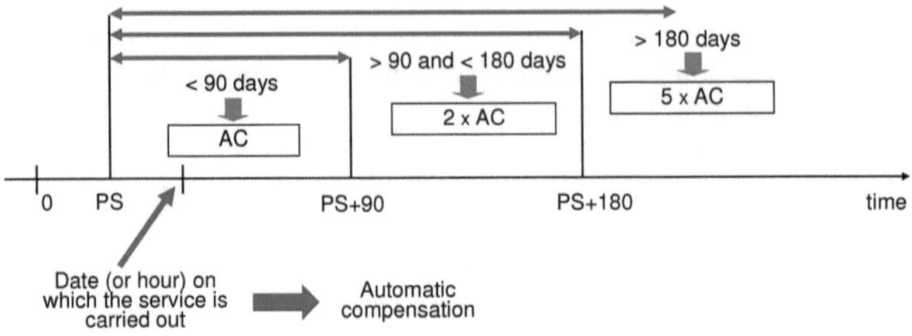


**Fig. 3.2.** Time-increasing automatic compensation

the required level of quality. For instance, a company is normally exempted from paying compensation if it was unable to gain access to the customer's premises, or if the customer has not paid its bill, or has provided incorrect information. Furthermore, on request compensation is normally not paid if the customer does not present a claim within a given period of time.

The regulator generally establishes that it is the responsibility of service providers to ensure that their customers are fully informed of their rights under the GS. Providers should take care to inform customers of those situations where automatic payment will be made for breaches of GS, and the procedures which customers need to follow in cases where a claim for compensation is necessary. These communications can be made by letter, or by means of hand-outs and posters at customer care centres, or they can be posted on the relevant website, and so forth.

### 3.3.2 Examples

The best way to illustrate the issues presented in the previous sections is to look at practical examples.

Two of the examples presented here are taken from the UK and the Hungarian commercial quality regulations. The UK regulation is a good example of a country in which distribution and retail activities have been legally unbundled. The Hungarian regulation represents a more traditional approach, where the distribution and the retail segments are only subject to accounting unbundling.

Finally, we present a small part of the Spanish regulation. The Spanish regulator has found a good solution for a difficult issue: classifying the requirements for estimating charges for connections to the distribution grid and for carrying out any related works.

#### 3.3.2.1 United Kingdom

The UK case is a very interesting example of an advanced commercial quality regulation. In the UK the local distribution network is operated as a regulated monopoly; by contrast, metering and retailing are competitive activities. The UK regulatory authority, Ofgem (Office of Gas and Electricity Markets), believes that is not necessary to set quality standards for market segments in which there is effective competition. Indeed, Ofgem considers effective competition to be the best protection, in terms of prices and services, for all customers.

According to this principle, Ofgem has progressively modified and then removed GS and/or OS where it considered competition to be sufficiently strong to ensure satisfactory commercial services and performance. It is interesting to note that, after the metering segment was opened to competition, Ofgem maintained MQS for the metering service for a period of time. These regulatory instruments continued to apply so long as Ofgem considered that the dominant companies were still in a position to exert control over this activity. Only when Ofgem was satisfied that competition in the metering markets had become *effective* were the MQS lifted [Ofgem, 2006c].

Table 3.3 lists the current GS applying to distribution companies (OS applying to the distribution segment were first introduced in 2001 and then removed in 2005). Note that there are a total of twelve GS; the empty lines in Table 3.3 correspond to GS that relate to continuity of supply [Ofgem, 2005a]. Note also that, for most GS, the payment is automatic, whereas in the case of services for which it is more

**Table 3.3.** GS in electricity (distribution), UK

| | Service | Performance standard | Compensation |
|---|---|---|---|
| 1 | Respond to failure of a distributor's fuse | 3 hours on a working day and within 4 hours on other days | £20 for domestic and non-domestic customers |
| ... | | | |
| 4 | Estimate of charges for connection | 5 working days for simple work and 15 working days for significant work | £40 for domestic and non-domestic customers |
| 5* | Notice of planned interruption to supply | Customers must be given at least 2 days notice | £20 for domestic and non-domestic customers |
| 6 | Investigation of voltage complaints | Visit customer's premises within 7 working days or provide an explanation of the probable reason for the complaint within 5 working days | £20 for domestic and non-domestic customers |
| 7 | Making and keeping appointments | Companies must offer and keep a morning or afternoon appointment, or a timed appointment if requested by the customer | £20 for domestic and non-domestic customers |
| 8 | Payments owed under the standards | Payment to be made within 10 working days | £20 for domestic and non-domestic customers |
| ... | | | |

*Compensation on request

difficult to identify who did not receive the required level of quality, compensation payments are paid on request (for commercial quality only the GS on 'notice of planned interruption of supply').

The UK regulation provides a detailed description of the services, obligations and exemptions indicated in the GS [Ofgem, 2006b]. A short version of this description is given below.

- *Response to failure of a distributor's fuse*: where a distributor is notified by a telephone call of an actual or apparent operation of its fuse so as to disconnect the supply, and the notification is received during working hours, an appropriate person must be sent to replace or reinstate the fuse and restore supply within 3 hours on working days and within 4 hours on any other day.
- *Estimating charges for connection*: once a customer has requested an estimate for the installation of a distributor's fuse and a service line for the delivery of a low voltage supply, and once the customer has given all information required, the distributor has to:
  - provide an estimate to the customer within 5 working days if there is no need for significant additional work other than the provision of a service line and distributor's fuse in order to establish the connection; or
  - provide such an estimate within 15 days if significant work is required in addition to that mentioned above.[5]
- *Notice of planned interruption of supply*: the distributor is required to give customers at least two days' notice in writing, when it discontinues supply for an authorized purpose, stating the day of the interruption. If the distributor fails to give the required notice, or if supply is interrupted on a day different from that notified, it must, on receipt of a qualifying claim, make a payment to the customer. Claims must be made within one month of the day on which the interruption occurred.
- *Investigation of voltage complaints*: if a customer reports that he believes that the supply is or has been outside the permitted voltage range, or if the customer reports an event which might reasonably lead the distributor to believe that this is the case, the distributor must either:
  - (where a visit is deemed necessary) make an offer to visit the customer's premises during a specified time within 7 working days; or
  - (where no such visit is deemed necessary) provide a written explanation within 5 working days of receipt of the customer's report.
- *Making and keeping appointments*: where the distributor informs a customer that it wishes to visit the customer's premises, or where a customer asks the distributor to visit its premises, the distributor must, within a reasonable period from the applicable date, offer a scheduled appointment. The appointment

[5] In UK regulation 'significant work' refers to a connection requiring work in addition to the provision of a service line and a distributor's fuse. In the UK moving a meter is a retailer responsibility and distributors may only do the work if so agreed by contract with the retailer.

offered must itself be within a reasonable period from the date when the request for the visit is made. It must be:
  - during a specified period up to 13:00 (the starting time of this period should also be notified to the customer) on a specified day;
  - during a specified period after 12:00 noon (the limits of this period must be notified to the customer) on a specified day; or
  - during a specified period not exceeding two hours in length on a specified day.
- *Payments owed under the GS system*: if a distributor is obliged to make a payment to one of its customers under any of the GS in the regulation it must make the payment to its customer or its customer's retailer within 10 working days of the date when the obligation to make the payment arises.

We refer to the Ofgem document (2006b) for further details on GS in the distribution sector. In particular, regulators should pay attention to the general and specific exemptions from GS. We report here an extract of the *general exemptions*, as an indication.

- If, in order to meet the guaranteed standard, information is required to be given by the customer to the distributor and the customer either sends the information to an address or telephones a number other than the one which the distributor has indicated, or (in the case of investigating voltage complaints) telephones with the information at a time outside reasonable hours as notified by the distributor.
- It was not reasonably practicable for the distributor to have complied with the requirements because of:
  - severe weather;
  - industrial action by the distributor's employees;
  - an inability to gain necessary access to relevant premises;
  - the likelihood that the distributor would break the law if he complied;
  - the effects of an event for which emergency regulations have been adopted under Part 2 of the Civil Contingencies Act 2004; or other exceptional circumstances beyond the control of the distributor,

  and the distributor in each case took all reasonable steps both to prevent the circumstances from occurring and from having that effect.
- It was reasonable for the distributor to regard information from the customer as being frivolous or vexatious.
- The customer had either committed an offence under the Electricity Act 1989 or had failed to pay charges due after receiving a disconnection notice.

As far as the metering activities are concerned, Table 3.4 reports the GS enforced until March 2007. As before, it is useful here to present an excerpt of the description of the regulated services [Ofgem, 2005a].

- *Response to meter problems*: where an electricity supplier is notified by a domestic customer that it considers that an appropriate meter is or may have been

**Table 3.4.** GS in electricity (metering), UK

| | Service | Performance standard | Compensation |
|---|---|---|---|
| 1 | Respond to meter problems | Visit customer's premises within 7 working days or provide an explanation of the probable reason for the complaint within 5 working days | £20 for domestic and non-domestic customers |
| 2 | Respond to pre-payment meter faults | Respond within 3 hours on a working day and within 4 hours on any other day | £20 for domestic and non-domestic customers |

operating outside the margins of error, or that there are circumstances which might reasonably have been caused by the meter operating outside the margins of error, the distributor must visit the customer's premises within 7 working days or provide an explanation of the probable reason for the complaint within 5 working days.
- *Response to pre-payment meter faults*: in the event that an electricity retailer is informed by a domestic customer who receives its supply through a pre-payment meter that the pre-payment meter is not operating so as to permit the customer's premises to be supplied, or if the retailer otherwise becomes aware of circumstances suggesting that it is not operating, an appropriate person must visit the premises where the pre-payment meter is installed in order to repair or replace it, within 3 hours on a working day or within 4 hours on any other day.

#### 3.3.2.2 Hungary

Hungarian commercial quality regulation has a more traditional structure. The obligations deriving from GS apply to distribution companies in their capacity as network operators, meter operators, and electricity retailers. However, in fixing OS the Hungarian Energy Office (HEO) has taken into account that the accounting unbundling regime currently in force separates the network (plus metering) from the retailing activity (no companies are yet legally unbundled).

The GS are listed in Table 3.5 [HEO, 2004, HEO, 2006]. Empty lines indicate that the GS concern either continuity of supply or voltage quality. In contrast to the UK case, the Hungarian regulation includes GS on 'responding to written requests for information' and 'correcting bills'. These services are normally regulated when there is no competition for retail activities.

An interesting feature of the Hungarian regulation is that for GS 4, 8, 10 and 13, compensation payments are automatic. For the other GS, companies are free to choose how to compensate customers: automatically or on request. Automatic payments are lower than on request payments (see Table 3.6). Obviously, this mechanism provides incentives for companies to pay compensation automatically. For GS 5 and GS 13, compensation is calculated as in Table 3.7.

**Table 3.5.** GS in electricity, Hungary

| | Service | Performance standard |
|---|---|---|
| ... | | |
| 3 | Estimating charges for connection | The company provides written information within 8 days in cases not requiring network intervention, within 30 days in cases requiring network intervention. |
| 4* | Providing connection | In case of a new connection or capacity increase the company connects the consumer within 8 working days (unless different different agreements are reached) |
| 5 | Making and keeping appointments | Upon a customer's request for a timed appointment, the company's representative arrives on the date and within the time frame agreed upon. The time frame cannot be longer than 4 hours. |
| 6 | Responding to written requests for information | The company answers within 15 days |
| 7 | Notice of supply interruptions | At least 4 days advance notice should be given for planned interruptions shorter than 4 hours and at least 8 days for interruptions longer than 4 hours. |
| 8* | Investigation of voltage complaints | The company contacts the customer within 10 working days with a proposal to adjust the voltage. Else the company proposes a date for installing a measuring device. Measures should beging within 5 days and the company informs the customer about the outcomes within 15 days after completing the measures. |
| ... | | |
| 10* | Correcting bills | The company refunds the consumer within 8 days after verifying the error in the electricity bill. |
| 11 | Responding to meter problems | Upon the request of the customer the company arranges for the meter to be examined, by simple means (such as counting the revolutions) within 15 days. The company replaces the defected, inaccurate, non-functioning meter within 8 days. |
| 12 | Reconnection following lack of payment | The company reconnects the customers on the day following the full and credible settlement of the debt, interests and all costs concerning the collection of the payment, in case of the existence of a valid contract. |
| 13* | Non lawful disconnections | The company pays a penalty in case it disconnect a consumer for no justified reason |

*Automatic compensation

**Table 3.6.** Compensation payments for GS 1 to 4 and 6 to 12

| | Domestic consumer | Non-domestic consumers | |
|---|---|---|---|
| Automatic | ~7 euros | Low voltage | ~21 euros |
| | | Medium voltage | ~54 euros |
| Upon request | ~18 euros | Low voltage | ~36 euros |
| | | Medium voltage | ~107 euros |

**Table 3.7.** Compensation payments for GS 5 and 13, Hungary

| | Domestic consumer | Non-domestic consumers | |
|---|---|---|---|
| Automatic | min. ~14 euros* | Low voltage | min. ~43 euros |
| | | Medium voltage | ~ 107 euros |
| Upon request | min. ~36 euros* | Low voltage | min. ~71 euros |
| | | Medium voltage | ~ 214 euros |

*The compensation is equal to the fee for an investigation made at the customer's premises. This fee is not regulated and thus can vary.

Hungarian OS are listed in Table 3.8. The overall performance of some companies is evaluated at company level. The regulator evaluates OS 1 to 4 and OS 6. These include network services ('connection of a new consumer' and 'meter readings') as well as activities that are relevant for both a distributor and a retailer. The performance of other companies is evaluated separately, for network services and retailing activities. In this case OS 5 is included and the performance standard is set at different levels for network activities and retailing activities.

#### 3.3.2.3 Spain

One of the delicate tasks in commercial quality regulation is the design of GS for 'estimating charges' and 'executing work' to connect a customer. The problem is that connections can require very simple to very complex interventions on the part of the network operator. Setting the boundaries between what can be considered a simple intervention and what should be defined as complex is not straightforward.

The Spanish regulator (Comisión National de Energía, CNE) has proposed a clear classification of the type of work that may be necessary for a new connection. Table 3.9 reports the classification for the 'estimating charges' service and the corresponding performance standards.[6] Table 3.10 reports analogous information

[6] In cases of non-compliance with performance standards, compensation is paid to customers in the form of a 10% discount on the first full bill (after connection has been made) up to a maximum of 30 euros. Without prejudice to that basic discount, the consumer may also claim compensation through the civil courts for any damages caused by non-compliance.

**Table 3.8.** OS in electricity, Hungary

| | Service | Performance standard |
|---|---|---|
| 1 | Responding to customers' requests | Domestic consumers: 90% within 7 days; 100% within 8 days<br>Non-domestic consumers 90% within 25 days; 100% within 30 days |
| 2 | Providing connection to new consumers | 90% within 7 days; 100% within 8 days |
| 3 | Responding to customers' requests (including complaints) | 90% within 12 days; 100% within 15 days |
| 4 | Speed of telephone response | 80% of calls must be answered within 30 seconds |
| 5 | Number of justified customer complaints received by the regulator | Different for different companies (between 0.014 and 0.93/ thousand consumers/year) |
| 6 | Number of meter readings | One per year |

for the 'execution of works'. As illustrated in these tables, performance standards depend on the voltage level and on the complexity of the connection [CNE, 2000].

GS on connections are also complicated by the fact that a number of external factors can delay the execution of works. Such factors include, for example, the time needed to obtain administrative authorizations for works related to the connection, or delays imposed by customers who have difficulties in finishing other works on the premises. These factors must be taken into account when measuring the time taken by the company to provide the service requested.

## 3.4 Reward and penalty schemes

Reward and penalty schemes are complex instruments that a regulator can employ to induce the regulated utility to deliver desirable levels of service quality. In general, reward and penalty schemes modify the company's revenues according to its actual performance as measured against performance standards set by the regulator. Such schemes are intended to replicate a competitive market mechanism, associating higher quality levels with higher revenues and vice versa.

Reward and penalty schemes are more difficult to design than GS and OS, and they require considerable work in the implementation phase. The use of this regulatory instrument is not widespread in commercial quality regulation. MQS seem to be a sufficient and effective means to improve the quality of commercial transactions. For these reasons it is advisable for regulators approaching commercial quality regulation for the first time to consider the use of reward and penalty schemes only after gaining some practical experience with MQS. By contrast,

**Table 3.9.** GS, estimating charges for connection, Spain

| Voltage level | Complexity | Performance standard |
|---|---|---|
| Low voltage | Load capacity up to 15 kW and network expansion is not needed | The distribution company shall give the technical-economic conditions in writing within 5 working days |
| | Any connection that does not require the installation of a transformer | The distribution company shall give the technical-economic conditions in writing within 5 working days |
| | Any connection that requires the installation of a transformer | Auxiliary works service: 10 days |
| | | Permanent service with medium to low voltage transformer centre: 20 working days |
| | | Permanence service with high to medium to low voltage transformer centre: 30 days |
| Medium voltage | Nominal voltage up to 66 kV | 40 days |
| | Higher voltages | 60 days |

reward and penalty schemes are the preferred instrument for regulating continuity of supply. For an extensive introduction to this topic, we refer the reader to Section 4.4. Here we simply report the two existing examples in Europe of reward and penalty schemes in the case of commercial quality regulation. They relate to the quality of call centres, and they are employed in the UK and in Ireland.

### 3.4.1 United Kingdom

The UK's reward and penalty scheme for commercial quality regulation applies to the quality of distributors' call centres. The incentive scheme uses two *quality indicators*: customer satisfaction with the speed of telephone response and quality of telephone responses [Ofgem, 2001b]. A customer survey results show that these indicators rank highest for customers.

As far as *data collection* is concerned, Ofgem (or its appointed consultants) undertakes a survey of the views of customers on the telephone response that they receive when they contact the distributor about a power loss or an emergency. To

**Table 3.10.** GS, execution of works, Spain

| Voltage level | Complexity | Performance standard |
|---|---|---|
| Low voltage | Network expansion is not needed | 5 working days |
| | Low voltage network expansion is necessary | 30 working days |
| | Installation of a transformer is necessary | 60 working days |
| | Installation of more than one transformer is necessary | 80 working days |
| Medium voltage | Mains* connection to a single customer with a nominal supply voltage up to 66 kV | 80 working days |
| | Higher voltages | Deadlines determined in each case according to the complexity of the work |

* In case the customer asks for several connections (normal/main and security/emergency connections), only the main one is concerned by the standard. Exclusively high voltage customers ask for several connections.

undertake this so called 'call-back survey', Ofgem requires that distribution companies register and report the telephone number of each person who has contacted the distributor by telephone (dialling the power outage telephone number and the security and safety enquiry service telephone number) together with (if known) the name of that person, whether that person is a domestic or non-domestic customer, and when they telephoned the distributor. Customers who have contacted the distributors are re-contacted and asked five questions concerning:

- satisfaction with speed of the telephone response;
- usefulness of information provided by the company;
- accuracy of information provided by the company;
- politeness of the relevant staff member;
- willingness of staff to help.

Responses are scored on a scale of 1 (very dissatisfied) to 5 (very satisfied). Responses to the first question measure an *annual mean performance* of the company with respect to the first indicator (satisfaction with the speed of telephone response). The other four questions are combined together to measure performance with respect to the second indicator (quality of telephone response). In particular, Ofgem calculates the *average performance score* across the four questions for each company, and the average across all fourteen companies (*industry average performance score*). In calculating the average performance score for a company, each of the four questions is weighted equally.

These actual measures of performance are used to calculate rewards and penalties. In practice, Ofgem calculates an *adjustment to* the company's allowed *revenue* (for the relevant year) using one method for satisfaction with the speed of response and another for quality of response.

As far as satisfaction with the speed of response is concerned, if the annual mean performance is between 4.1 and 4.5 there are no financial consequences for a company (this interval is called a 'dead-band'). For a performance of between 3.6 and 4.1, companies are subject to a sliding-scale penalty (in the form of a negative adjustment to revenue that is incrementally greater for poorer performance). For a performance equal to or lower than 3.6 the negative adjustment to revenue no longer increases: the penalties are capped at 0.25% of revenue (the revenue linked to the speed of telephone response). For a performance higher than 4.5 there is a small reward in the form of a 0.05% positive adjustment to revenue.

As for the quality of telephone response, the adjustment to revenue for each company is calculated as follows:

- the standard error is added to the average performance score to obtain the *deemed performance score*;
- the company-specific *incentive rates* are calculated:
  - (for companies whose performance is below the industry average performance score) by dividing the amount of revenue that each company has linked to the quality of telephone response indicator (0.125% of revenue) by the difference between the industry average performance score and the lowest deemed performance score, i.e., the score of the lowest ranked company; or
  - (for companies whose performance is above the industry average performance score) by dividing the amount of revenue that each company has linked to the quality of telephone response indicator (0.125% of revenue) by the difference between the industry average performance score and the highest deemed performance score, i.e., the score of the highest ranked company;
- the adjustment to revenue is calculated for each company by multiplying the incentive rate by the difference between a company's deemed performance and the industry average performance score.

The calculation of the adjustment to revenue is shown in Table 3.11. The revenues of a company are adjusted for an amount that depends on the distance between the actual, measured performance of the company and the performance standard. This distance is multiplied by the incentive rate (in £m per unit of performance score) to obtain the adjustment to revenue (note that the incentive rate is different for each company). It is important to note that this revenue adjustment is also capped, and thus has a lower and an upper limit. The adjustment cannot be greater than the amount of revenues that each company has linked to the quality of telephone response.

**Table 3.11.** Quality of telephone responses: a numerical example, UK

| | | | B | C | D=C/(Y-A) or D=C/(A-X) | E=B-A | E·D |
|---|---|---|---|---|---|---|---|
| Company | Average performance score | Standard error | Deemed performance score | Amount of revenue exposed (£m) 0.125% | Incentive rate (£m per unit of score) | Difference between deemed and industry average score | Adjustment to revenue (£m) |
| 1 | 4.3 | 0.2 | 4.5 **(Y)** | 0.3 | 0.5 | 0.65 | +0.3 |
| 2 | 4.1 | 0.15 | 4.25 | 0.2 | 0.3 | 0.4 | +0.1 |
| Industry average | 3.85 **(A)** | | | | | | |
| 3 | 3.5 | 0.25 | 3.75 | 0.2 | 0.8 | -0.1 | -0.1 |
| 4 | 3.4 | 0.2 | 3.6 **(X)** | 0.2 | 0.8 | -0.25 | -0.2 |

### 3.4.2 Ireland

As in the case of UK, the Irish regulatory authority (Commission for Energy Regulation, CER) employs a reward and penalty scheme for the regulation of the quality of call centres. The regulation applies to the *distribution system operator* (ESB DSO, Electricity Supply Board Distribution System Operator) and to the *dominant retailer* (ESB PES, Electricity Supply Board Public Electricity Supplier) [CER, 2006]. A legal unbundling regime applies to ESB, the vertically integrated, leading company in the electricity sector in Ireland. The four *quality indicators* in the incentive mechanism are:

- speed of telephone response
- call abandonment rate
- customer call-back survey results
- mystery caller survey results

*Data collection* regards the level of service provided by ESB's National Customer Contact Centre (NCCC). In general all customer contacts are either about ESB's retailing activity (PES) or about the distribution network operated by the DSO. Customers of other retailers in the market contact the NCCC about network-related issues.

The speed of telephone response and call abandonment rate are conventional measures of call centre performance. Software was installed that provides for the separate recording of retail and network queries. Two indicators are used to evaluate the *speed of response:*

- the percentage of calls to the call centre answered (either by a member of staff or a recorded voice) within 20 seconds (referred to as STP1);
- the percentage of calls in a queue waiting to speak to an agent (after being placed in a queue either via the recorded voice or by an agent) that are answered by an agent within 30 seconds (referred to as STP2).

These indicators are combined and expressed as a percentage, with a score of 100 representing a perfect rating.

The indicator for *call abandonment rate* records the number of calls that are abandoned while a caller is waiting in a queue to speak to an agent.

The customer call-back and mystery caller surveys are conducted by an independent organization approved by the regulator and funded by PES and DSO. The sample for the customer *call-back survey* includes both network and PES calls. The calls are selected randomly, subject to the (reasonable) inclusion of calls by time of day when the call was made, the purpose of the call (e.g. bill query, supply problem, meter reading), and handling of the call (on-call resolution, requiring call-back, requiring referral). Customers are asked to score their call centre experience on a scale of 1 (very dissatisfied) to 5 (very satisfied) on:

- politeness of the staff member
- their willingness to help

- the accuracy of any information given
- the usefulness of any information given

An overall performance score is calculated for this indicator, expressed as a percentage, with a score of 100 representing a perfect rating.

The *mystery caller* survey involves a third party acting as a genuine caller, making calls to assess various aspects of customer services. Aspects of the call centre agent's approach and disposition are evaluated, including helpfulness, responsiveness, tone and style of the agent. Aspects of information delivered are also rated, including correctness, completeness, coherence and pace of the call. Each call made is scored and an overall performance score is calculated for this indicator. This is expressed as a percentage, with a score of 100 representing a perfect rating.

Finally, each quality indicator is multiplied by a weighted figure to produce a *comprehensive* quality of service performance. The weighting of the four indicators is linked to what is important for customers. For instance, the time taken to answer calls is weighted more heavily than the call-back survey measure as the regulator believes an improved performance on the speed of telephone response is currently more important to customers (see Table 3.12).

The actual, measured performance is set against a *comprehensive performance standard.* The objective of the regulatory scheme is explicitly to provide incentives to reach a higher level of performance. The comprehensive performance standard becomes more demanding over the 5-year period of the regulation (see Table 3.12).[7]

Financial rewards and penalties will depend on the distance between the measured, comprehensive performance and the comprehensive performance standard

**Table 3.12.** Performance standards and weightings, Ireland

| Year | | 2006 | 2007 | 2008 | 2009 | 2010 |
|---|---|---|---|---|---|---|
| | Weights | Performance std. | | | | |
| Speed of Tel Response (STP1 and STP2) | 30% | 70% | 75% | 80% | 83% | 83% |
| Abandonment Rate | 30% | 5% | 5% | 5% | 5% | 5% |
| Mystery Caller | 20% | 67% | 72% | 75% | 80% | 80% |
| Call-back Survey | 20% | 77% | 78% | 80% | 80% | 80% |
| | **100%** | | | | | |
| **Comprehensive performance std.** | | **78%** | **81%** | **84%** | **85%** | **85%** |

[7] Performance standards for the speed of response are derived from two separate standards (not reported here) for the STP1 and STP2 measures that are combined together, assigning equal weight to the two measures.

**Table 3.13.** Financial rewards and penalties, Ireland

| Year | | 2006 |
|---|---|---|
| Incentive rate (Euros) | | 95000 |
| Allowed Revenue (mEuros) | | 72000000 |
| Max. Penalty (mEuros) | 1.00% | 720000 |
| Max. Reward (mEuros) | 0.25% | 180000 |
| Performance at which max penalty paid | | 71% |
| Performance at which max reward paid | | 80% |

for the relevant year. The *incentive rate* - defined as the payment to be made to, or by, the company per 1 percentage point deviation between actual performance and the performance standard - is calculated for each year by the regulator. The value for the year 2006 is indicated in Table 3.13. If the comprehensive performance standard is 78% for the year 2006, a measured, comprehensive performance of 80% will mean a reward of twice the incentive rate. The incentive scheme is capped. The maximum penalty is 1% of the company's allowed revenue (for the relevant year). The maximum reward is 0.25% of the company's allowed revenue. Rewards thus have a lower cap than penalties. Table 3.13 summarizes this information for the year 2006.

## 3.5 Advanced topics

The opening up of electricity markets to full retail competition challenges the traditional approach to commercial quality regulation. On the one hand, regulation of commercial transactions that occur between customers and the network operator is generally considered necessary. On the other hand, at least in principle, activities that have been opened up to competition do not need to be regulated. Customers should be able to rely on the protection afforded by competition. However, before removing the regulatory instruments that are already employed at least two factors should be considered.

As we have seen in the UK example, an activity (such as metering) can be open to competition and at the same time it may still be controlled by a dominant operator. In other words, there may be a degree of competition, but it may not be considered sufficient to justify the removal of MQS. Moreover, the reference supplier for a number of transactions could be the retailer (i.e., an operator in the open market), but in practice the service may be carried out by a monopolist. This is true, for instance, for the reconnection of customers following disconnection for debt. In this case, the regulator may require that retailers have contractual arrangements in place to ensure that the network operator makes the reconnection within the expected time.

Taking into consideration both the general principle and the complexities of the real world, a regulator can, thus, consider different options:

- setting regulatory performance standards for all suppliers;
- setting regulatory performance standards for suppliers that are dominant in the electricity market in order to protect customers who have not taken advantage of the competitive market;
- removing the existing performance standards and relying on the competitive market;
- allowing suppliers to set their own performance standards in codes of practice.

As regulations in EU member states are revised and modified to take into account full retail competition, more experience will become available. Newcomers in the field should benefit from the observation and analysis of the outcomes of the different choices made.

# 4

# Continuity of supply

Continuity of supply relates to the interruption of service to customers connected to the distribution network.

In practice, continuity of supply regulation uses all four regulatory instruments described in Chapter 2. The basic prerequisite for any regulatory policy is the availability of reliable measures of relevant quality dimensions. Hence, regulatory instructions and guidance on data collection will be described first (Section 4.1). A detailed description of the regulatory instruments follows: publication of performance data in Section 4.2, MQS in Section 4.3, reward and penalty schemes in Section 4.4, and premium quality contracts in Section 4.5. Section 4.6 presents a few advanced topics.

There is no reason to design a regulatory policy that includes all four regulatory instruments, especially at the outset. On the contrary, it is better to introduce one instrument at a time (beginning with data publication), and to gradually increase their number (or the number of quality indicators that are regulated with each instrument). Several practical tasks (some of them quite complex and time-consuming) contribute to the efficient design and implementation of each instrument. All of them should be examined, and a lot of fine-tuning of the regulatory design can and should be carried out over time (in subsequent revisions of the applied mechanisms). Experience suggests that the introduction of continuity regulation can take one to two years' work (or three to four if reliable data are not available) on the part of both the regulator and the distribution companies.

## 4.1 Definitions and data collection

Continuity of supply concerns a single service, the *supply of electricity to customers.* The main quality dimension of this service is its continuity or, in other words, the absence of interruptions. Interruptions in the supply to final customers depend on the reliability of the entire power system, made up of a generation system, a transmission grid, and a distribution network. However, as explained below (Section 4.1.1), this handbook is concerned mainly with the reliability of the distribution network. Clearly, in a regime of accounting or legal unbundling of distribution and retail services, continuity of supply regulation will only apply to the *network operator*.

Our perspective coincides with that of final customers. From the perspective of a network user, the disturbance caused by interruptions in electricity supply is well described by the number of events in a period of time and the duration of service interruptions.[1] This is why continuity of supply regulation focuses on *quality indicators* that capture the frequency and duration of supply interruptions. Failure rates of system components and similar indicators are the concern of the distribution company, but are only intermediate measures of the quality dimensions that customers value directly.

Thus, the objective of regulatory instructions and guidance on data collection is to gather reliable information that will describe the performance of the distribution network with respect to the number and duration of supply interruptions. A few general features of regulatory instructions for data collection are the following:

- information must be collected in a uniform manner by all distribution companies to ensure fairness;
- procedures for measuring quality indicators must be clearly defined in order to audit the collected data;
- some indicators will be subject to regulation: the information recorded must be consistent with the objective of the regulatory instruments to be introduced;
- any information recorded must be consistent with the technical equipment available. It is normally easier to collect information on the average performance of the utility (over the distribution territory or a part of it) than to collect information on interruptions at individual delivery points. Hence, it is advisable to begin with *system level measurements* and then move to individual ones. Similarly, it is easier to keep a record of the number and duration of *long interruptions* (over three minutes) than to measure the number of *short interruptions* (up to three minutes).[2] Accordingly, this handbook attaches higher priority to the collection of data on long interruptions at system level. Regulatory policies should focus on these measurements for the first years of implementation. Indeed, long interruptions are the main concern of domestic customers.

The process of data collection on continuity of supply can be broken down into four main steps: *registering* all interruptions and their characteristics; computing the relevant *quality indicators* from the register; *reporting* these indicators to the regulatory authority (or to individual customers); and *verifying* the reliability of the figures reported by the companies concerned. The regulatory instructions should

[1] The 'duration' of service interruptions in fact comprises two causes for dissatisfaction: the duration of each interruption and the total amount of time in a given period (for instance, a year) that the consumer is without supply. We will specify which concept is being referred to whenever necessary.

[2] In this handbook we adopt the definitions given in the European Standard 50160 [CENELEC, 1999].

define rules for all these tasks. Section 4.1.2 describes these rules for system-level measurement of long interruptions (with the help of a simple example), Section 4.1.3 describes individual measurement of long interruptions, and Section 4.1.4 deals with measurement of short interruptions.

Before entering into further detail, it is necessary to focus on an important point. Writing a regulatory protocol for companies to follow when measuring and registering data on continuity of supply means that the regulator must have a thorough understanding of the physical events that are being recorded as well as of the basic functions performed by the equipment of a distribution network. To this end, Section 4.1.1 provides a brief introduction to the subject, but for obvious reasons it cannot be exhaustive. The interested reader should refer to the vast engineering literature on network reliability and system protection and control.

### 4.1.1 Interruption events

Customers are no longer supplied with electricity when there is an outage in the supply, i.e., the removal of a primary component from the power system. The main focus of continuity of supply regulation is on distribution component outages. Indeed, distribution systems are often radially operated (without redundancy), especially at lower voltage levels. Thus, each component outage will lead to a supply interruption. By contrast, generation and transmission systems have a much higher degree of redundancy. As a consequence, generation outages and outages of transmission components account for a very small percentage of interruptions at the distribution level. In addition, a distribution network operator has no control over the reliability of the generation or transmission system which are owned and operated by different companies.

It is therefore important to identify which system component is at the *origin* of the supply interruption. Exemptions from the payment of compensation (in the case of GS) or from the calculation of penalties and rewards will usually apply for interruptions caused by events originating in the generation system and the transmission network. In contrast, with some exceptions, the regulator holds the distribution company responsible for events originating in the distribution system. In addition, understanding where most of the outages occur on the distribution system enables a closer monitoring of the system performance.

There is a significant difference between planned and unplanned interruptions (interruption *type*).

Planned interruptions normally occur when a component is removed from the system to allow scheduled maintenance work. This will not affect customers if the network system has some level of redundancy. However, at lower voltages redundancy is often not available, which means that preventive maintenance and repair or changes in the system can only be performed when the supply to a portion of customers is interrupted. Customers should be informed of any planned interruption well in advance. Regulators often use the terms 'notified' and 'un-notified' interruptions: for the customer, what is important is not whether the interruption is planned or unplanned, but whether or not they are notified about it a reasonable time in advance. The notification allows them to take some precautions and to mitigate the consequences of the interruption.

Long, unplanned interruptions are caused by component outages that arise when a fault occurs in the power system, leading to an intervention by the protection system that removes the faulted component from the network.[3] If the fault occurs in a part of the system that is not redundant, the intervention of the protection system leads to an interruption for a number of customers. This is a correct intervention of the protection system. If the protection does not intervene, the fault will probably lead to an interruption for a much larger group of customers and cause serious damage to electrical equipment [Bollen, 2000].

A fault can occur for a number of different reasons. There are equipment failures (due to deterioration from age and wear, to insufficient maintenance, incorrect settings of technical parameters, faulty manufacturing, and so on), weather-related events (lightning, snow, ice, etc.), environment-related events (falling trees, animals), and third-party interference (accidental damage or contact by other service operators, by their contractors, or by customers themselves). For instance, Ofgem (2000a) reports that approximately 50% of high voltage overhead line faults are related to weather conditions, 10% to deterioration due to age and wear, and 20% to unknown causes. Many faults are due to 'unknown' reasons due to a genuine difficulty in identifying a direct cause, particularly when there is no damage and the circuit is restored without the fault being found. Regarding underground cables, Ofgem (2000a) indicates that 10–30% of faults are caused by third-party damage, mainly due to roadworks. It is not practical to examine all faults in detail and companies reserve such analysis for situations where they have detected a specific problem. In any case, a regulator will not normally require companies to register this sort of information.

Alternatively, the regulator often defines a few 'regulatory *causes*' (two or three) which, if verified, lead to an exemption from the payment of financial penalties. The idea is to identify as clearly as possible those events that are not under the control of the company (often called 'exceptional events'). Several variables help convey the idea of such an event and are used differently in different countries. In general they include severe weather conditions (exceeding the technical design limits of the network, or leading to protracted and widespread interruptions) and non-weather conditions, such as public authority (police, firemen) interventions, strikes, and unavoidable damage caused by third parties such as, a plane crash, a terrorist attack, and similar events. It is not uncommon to group them all under the name of events caused by *force majeure*, a concept described earlier in this handbook (see, e.g., Section 3.1). Once the causes of exemptions are identified, all the events that are not related to them will be subject to financial penalties (including interruptions for unknown reasons). The subject of exemptions is particularly important in everyday practice and is discussed further in Section 4.6.

[3] Long, unplanned interruptions can also occur when the protection system intervenes incorrectly, thus causing a component outage [Bollen, 2000].

In some countries the regulator does not define 'regulatory causes' of exemptions but will consider excluding particular events at the request of the company. Many factors help shape the different approaches, including the prevalent administrative and legal framework of the country. In 'public law' countries a legal definition of events that can be excluded is usually given. In 'common law' countries it is more common for events to be analyzed on a case-by-case basis.

As noted above in Section 4.1, in terms of *duration*, unplanned interruptions are classified as 'long' when they last longer than three minutes, and as 'short' when they last for up to three minutes [CENELEC, 1999]. The duration of a long interruption depends on the time necessary to: become aware of the interruption, locate the fault, perform switching actions to a healthy supply or repair/replace the faulted components, and restore supply. Companies will always be required to register the starting and ending times of interruptions. The duration can be calculated from these two pieces of information. It is important to note that, in the absence of a control system on the network (see Section 4.1.2.4), the starting time of a long interruption will often coincide with receiving an alarm from a customer reporting the lack of supply. As a consequence, the duration computed from the register will be only an approximation of the actual duration of the interruption. This will certainly be the case for interruptions which originate in the lower voltage parts of the network.

### 4.1.2 Long interruptions: system-level

When deciding what companies should measure and report to the regulator, it is important to bear in mind exactly what type of quality indicators one is interested in. In particular, the regulation of long interruptions at system level requires indicators of the *average performance* of the company across the distribution territory (or a portion of it) during a specified period of time, known as the reporting period (normally one year). In practice, regulators can use several available statistical indicators to measure this performance.[4] We focus on the two used most frequently: the 'average number of interruptions per consumer per year' and the 'average interruption duration per customer in minutes per year' (defined in Section 4.1.2.2). However, a non-negligible number of regulators prefer to use, as an indicator of the average performance, the 'energy-not-supplied' (also defined in Section 4.1.2.2). The reader should be aware that the indications given in this section have a general validity, but they are tailored in particular to obtaining reliable measures of the former two quality indicators.

#### 4.1.2.1 The register

There are two main aspects to bear in mind when introducing a protocol for the registration of long interruptions at system level. The first is to make sure that the regulated company *keeps track of all the interruption events*. The second is to make

[4] For a complete reference see the IEEE Standard 1366–2003 [IEEE, 2004].

sure that the company *keeps track of all the relevant information* regarding each interruption event.

As far as the first aspect is concerned, regulators should introduce an obligation for regulated companies to annotate each interruption event chronologically, as they occur, in an appropriate register.

Concerning the second aspect, regulatory instructions must indicate clearly what type of information companies should register for each event. The main objective of the request for information is to acquire a better understanding of the level of supply continuity offered to customers. Details on the interruption event are also necessary to apply the regulatory instruments correctly.

In practice, companies should register five fundamental characteristics of the interruption event:

- *origin* of interruption: e.g. the transmission network or distribution network; in the latter case, it may help to register the name of the faulted component and the voltage level (high, medium or low voltage);[5]
- *type* of interruption (planned or unplanned);
- 'regulatory' *cause* of interruption (i.e. if any exemption applies);
- *classification* of interruption in terms of its duration (long or short);
- *number of customers affected* by the interruption.

The first four characteristics were briefly discussed in Section 4.1.1.

Computing the number of customers affected by an interruption requires a study of customer records which is often time-consuming. When feasible, the number of customers involved in each interruption may be identified from a *connectivity model* which links customer information to the section of the network where they are connected. It is important for the connectivity model to be kept up to date. The accuracy with which the number of customers affected is reported is of course partly determined by how often the connectivity model is updated [Ofgem, 2005b]. When a connectivity model is not in place, one can assume that the number of customers connected to each feeder remains constant over the reporting period. This number may be set at the beginning of the year and kept constant, disregarding changes in network configuration or in connections and disconnections.

At this point it may be useful to introduce a sample distribution grid and to use it to illustrate the concepts introduced regarding the register, the calculation of the statistical indicators (in Section 4.1.2.2), and the reporting requirements (in Section 4.1.2.3).

[5] According to the EN 50160, 'low voltage' is a voltage used for the supply of electricity whose upper limit of nominal rms value is 1 kV, and 'medium voltage' is a voltage whose nominal rms value lies between 1 kV and 35 kV [CENELEC, 1999]. The definition of high, medium and low voltage portions of the distribution network can differ across countries.

**Example**

Consider the sample distribution grid in Figure 4.1. This simplified system has three different voltage levels: high voltage (HV), medium voltage (MV), and low voltage (LV). There are two HV/MV transformer stations from which medium voltage feeders depart. On the MV lines each circle represents a MV/LV substation. From these substations, LV feeders depart (see the detail on the lower left inside). Circuit-breakers (represented as squares) are located upstream and downstream of each transformer and at the departure of each feeder. The grid is radially operated, but each substation can be fed from either of the two HV/MV transformers, changing the position of the open switch (vertical bar). The MV feeder in the upper part of the figure has the highest density of MV/LV transformers and it serves an urban area. The central feeder goes through a sub-urban area, and the lowest one goes through a rural area.

A sample register for this grid is composed of two parts, illustrated in Tables 4.1 and 4.2.

In Table 4.1 the utility keeps the main data regarding the distribution territory: the number of customers in the urban area (U), in the suburban (S), and in the rural (R) areas, the number of transformers in each area, and the consumers/transformer ratio (cons/transf). These data are relevant because they will be used in computing the statistical indicators (Section 4.1.2.2). For the sake of simplicity, the regulator can allow companies to keep these data fixed for the entire reporting period (normally one year).

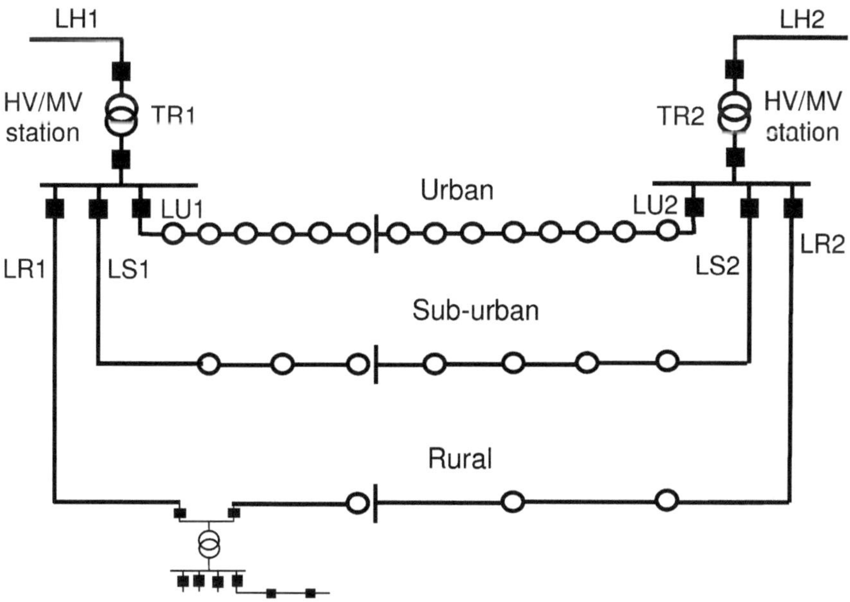


**Fig. 4.1.** Sample distribution grid

**Table 4.1.** Main data

| Sample Grid | Consumers | Transformers | cons/transf |
|---|---|---|---|
| U | 7000 | 14 | 500 |
| S | 1400 | 7 | 200 |
| R | 200 | 4 | 50 |
| Total | 8600 | 25 | / |

Table 4.2 gives a chronological list of interruptions: each line corresponds to an interruption event. For each event, the utility registers: the origin (all events in Table 4.2 are distribution system events and the voltage level where the outage occurred is given); the faulted component; the cause of exemption (events due to *force majeure*, FM), when applicable; the date, the starting and ending time of the event; the interruption type and its classification (planned or unplanned, long or short); and the number of customers affected.

The definition of duration of the interruption includes some complicated cases. For instance, a temporary restoration of supply which does not exceed three minutes can be ignored, and the interruption can be registered as a single long interruption. Definitions for this and similar cases can differ across countries but must be indicated in the measurement protocol.[6]

As far as the number of customers is concerned, we assume that a connectivity model is not in place. An acceptable approximation of the number of customers per feeder can be obtained by the consumers/transformer ratio. Note that the total number of customers served in the rural area in Figure 4.1 is 200: having 4 MV/LV transformers in the area, on average 50 customers are connected to each transformer. If 4 feeders depart from each transformer, the average number of customers per LV rural (LVr) feeder is 12.5.

Consider the first event in Table 4.2, as illustrated in Figure 4.2. A fault on the HV side of the HV/MV station on the left inside disconnects a total of 3,700 customers: the number of customers affected are 3,000 in the urban area (6 transformers serving 500 customers each), 600 in the sub-urban area, and 100 in the rural area.

The second event in Table 4.2 concerns the tripping of the breaker at the beginning of the urban MV feeder (component LU2). The number of affected customers is 4,000 in total (8 transformers serving 500 customers each).

Note that two entries are registered for the third event (see Tables 4.2 and 4.3). This is an example of step restoration: multiple restoration steps are required to restore service to all customers. As illustrated in Figure 4.3, the interruption has the same starting time for a total of 100 customers; all customers remain disconnected for 30 minutes; then, 50 of them are restored and another 50 remain disconnected for another 1 hour and 30 minutes. As explained below, this will introduce a small complication in the computation of the statistical indicators.

[6] On this topic, see also Section 4.1.4.

**Table 4.2.** Interruption register

| | Origin | Component | Exemption | Date | Start time | End time | Type | Affected consumers |
|---|---|---|---|---|---|---|---|---|
| 1 | HV | LH1 | yes: FM | Feb 3rd | 8.00 am | 8.15 am | Unp. long | 3000 U; 600 S; 100 R |
| 2 | MV | LU2 | yes: FM | Mar 12th | 4.05 pm | 5.45 pm | Unp. long | 4000 U |
| 3.1 | MV | LR1 | no | May 1st | 4.30 am | 5.00 am | Unp. long | 50 R |
| 3.2 | MV | LR1 | no | May 1st | 4.30 am | 6.30 am | Unp. long | 50 R |
| 4 | LV | 2 LVr | no | Jun 10th | 10.00 am | 11.40 am | Unp. long | 25 R |
| 5 | MV | LR1 | no | Oct 27th | 9.00 pm | 10.00 pm | Unp. long | 100 R |

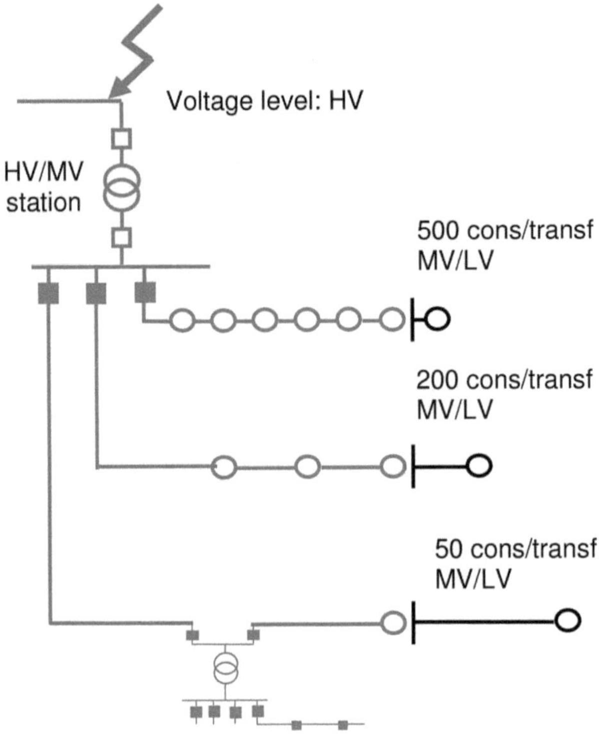


**Fig. 4.2.** HV interruption event

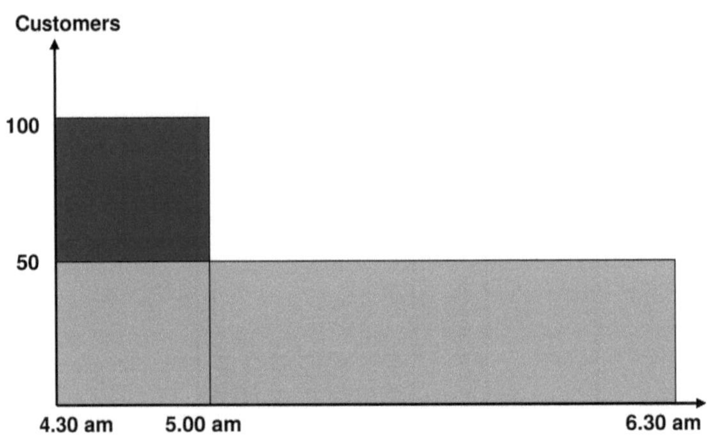


**Fig. 4.3.** Step restoration

#### 4.1.2.2 Statistical indicators

The most frequently used statistical indicators of performance are:

- average number of interruptions per customer per year;
- average interruption duration per customer per year; or
- energy-not-supplied.

We begin by defining of the statistical indicators, and then we discuss how they are computed in practice (focusing on the first two indicators).

Consider a utility serving $N_{tot}$ customers. During the reporting period (normally one year) a total of $K$ outages in the system lead to a long interruption affecting one or more customers. Interruption $i$ affects $N_i$ customers and has a duration of $D_i$ minutes.

The average number of interruptions per customer per year (internationally known as the System Average Interruption Frequency Index, SAIFI) is given by:

$$SAIFI = \frac{\sum_{i=1}^{K} N_i}{N_{tot}}$$

The SAIFI indicator indicates how often the average customer experiences a long interruption over a predefined period of time (a year). It should be noted that system design and operation as well as weather patterns can vary significantly within the same distribution territory. Hence, not all customers in the distribution territory will experience the number of interruptions indicated by the SAIFI indicator. Calculating this statistical indicator on smaller, more uniform portions of the distribution territory is thus advisable. The typical geographical separation is between urban and rural areas which differ in terms of customers density (higher in urban areas than in rural ones), average length of distribution feeders (higher in rural areas than in urban ones), and proportion of the network above ground (higher in rural areas than in urban).

The average interruption duration per customer in minutes per year (System Average Interruption Duration Index, SAIDI) is calculated as:

$$SAIDI = \frac{\sum_{i=1}^{K} N_i D_i}{N_{tot}}$$

The SAIDI indicator indicates the total duration of interruption for the average customer during a predefined period of time (a year). The same comments regarding geographical differences apply with respect to the SAIFI.

The other statistical indicator that is frequently used is the Energy-Not-Supplied (ENS). Indicating with $P_i$ the capacity disconnected during interruption $i$, the ENS in kWh per year is given by:

$$ENS = \sum_{i=1}^{K} P_i \cdot D_i$$

Turning to discuss how these indicators are calculated in practice, the SAIFI and the SAIDI are better computed for each interruption event, using the information already recorded in the register. It is helpful to expand the register in Table 4.2 to include the statistical indicators. It should be noted that, in Table 4.3, there are three entries, one each for urban, sub-urban and rural areas.

Consider the first event in the register: it had a duration of 15 minutes and affected 3,000 urban customers. Given that the utility serves a total of 7,000 urban customers:

$$SAIDI_U = \frac{15 \cdot 3000}{7000} = 6.4\,\text{min}$$

Analogously for the rural area:

$$SAIDI_R = \frac{15 \cdot 100}{200} = 7.5\,\text{min}$$

Sub-urban indicators are calculated in a similar manner, but for the sake of brevity they are not illustrated here.

Proceeding in the same manner, the SAIFI for the urban area is:

$$SAIFI_U = \frac{3000}{7000} = 0.4\,\text{interruptions}$$

And for the rural area:

$$SAIFI_R = \frac{100}{200} = 0.5\,\text{interruptions}$$

For events leading to a step restoration a criterion is needed for companies to calculate the statistical indicators in a uniform manner. As illustrated in Figure 4.3, using the information in rows 3.1 and 3.2, Table 4.2, we can calculate the SAIDI for 'two events': one affecting 50 customers for 30 minutes and one affecting 50 customers for 120 minutes:

$$SAIDI_R^{step1} = \frac{30 \cdot 50}{200} = 7.5\,\text{min}$$
$$SAIDI_R^{step2} = \frac{120 \cdot 50}{200} = 30\,\text{min}$$

**Table 4.3.** Interruption register and statistical indicators

| | ... | $SAIDI_U$ | $SAIDI_S$ | $SAIDI_R$ | $SAIFI_U$ | $SAIFI_S$ | $SAIFI_R$ |
|---|---|---|---|---|---|---|---|
| 1 | ... | 6.4 min | 6.4 min | 7.5 min | 0.43 | 0.43 | 0.50 |
| 2 | ... | 57.1 min | / | / | 0.57 | / | / |
| 3.1 | ... | / | / | 7.5 min | / | / | 0.25 |
| 3.2 | ... | / | / | 30.0 min | / | / | 0.25 |
| 4 | ... | / | / | 12.5 min | / | / | 0.12 |
| 5 | ... | / | / | 30.0 min | / | / | 0.50 |

Similarly, we can calculate the SAIFI for 'two events', each of which affects 50 customers:

$$SAIFI_R^{step1} = \frac{50}{200} = 0.25 \, \text{interruptions}$$
$$SAIFI_R^{step2} = \frac{50}{200} = 0.25 \, \text{interruptions}$$

Hence, the third event in the register contributes a total of 37.5 minutes to the annual rural SAIDI indicator.

In the same manner, this event contributes a total of 0.5 interruptions (i.e. the SAIFI of one event affecting a total of 100 customers out of 200) to the annual rural SAIFI.

The reader can verify the remaining figures in Table 4.3.

Regarding the ENS indicator, it is important to give at least a general idea about how it is calculated in practice. Computing $ENS_i$ for an interruption event $i$ that affected a distribution area requires detailed information on: ($i$) the starting time and ending time of the interruption, ($ii$) the number and type of customers affected; and ($iii$) an estimate of the capacity that each customer would have withdrawn from the network if the interruption had not occurred. This last figure will be derived from the customer load curve. It will be either a load curve defined for that particular customer or a standardized load curve related to the particular customer group to which the customer belongs.[7] This means that $P_i$ (the capacity disconnected during interruption $i$) will not be a constant value throughout the duration of the interruption. In fact, the energy-not-supplied to a customer is the integral, over the duration of the interruption, of the capacity indicated in the customers load curve. $ENS_i$ for the interruption event $i$ will be the sum of the energy-not-supplied to all the affected customers in the area.

In practice, the estimate of the ENS indicator becomes more precise as the necessary information is given with greater accuracy. For instance, the estimate of the capacity that would have been withdrawn from the network can be greatly improved by increasing the number of customer groups and the accuracy in the load profiling for each group.

#### 4.1.2.3 Report

Statistical indicators of performance are communicated to the regulator on a regular basis (normally once per year). The content, timing and form of these communications are set by the regulator.

The annual statistical indicators can easily be computed from the register. For regulatory purposes it may be necessary to break down the data in a specific manner. In particular, it may be necessary for the regulator to receive separate statistical

[7] Standardized load curves for different customer groups may be developed through research projects. Individual load curves may be developed through the use of hourly based measuring campaigns (these should last at least one year).

indicators of performance for different areas of the distribution territory (urban, sub-urban and rural). This is relevant when different performance standards are employed for different geographic areas. In addition, the regulator must be able to identify the statistical figures regarding the events to which exemptions apply.

Two sample report forms are illustrated in Tables 4.4 and 4.5 for SAIDI and SAIFI respectively.

These tables contain annual SAIDI and SAIFI indicators for the areas Urban and Rural in the previous sample network (for the sake of brevity Sub-urban indicators are not illustrated here): a distinction is made regarding the origin (transmission grid, different voltage levels on the distribution network) as well as the regulatory causes of interruption.

The annual values are simply the sums of the statistical indicators computed for all events with the same characteristics. For instance, the 67.5 annual SAIDI in the rural area, due to interruptions originated on the MV, and net of exemptions, result from events 3.1, 3.2 and 5 in the register.[8]

The forms in Tables 4.4 and 4.5 are for unplanned interruptions. In practice, companies also report similar forms also for planned interruptions.

The figures in the report run to 15 for each table. This detailed information helps regulators to identify the actual strengths and weaknesses of the distribution system and, over time, to evaluate the results of the regulatory policy in detail. As will be better explained later, the regulatory instruments will link the allowed revenues of the company to a subset of the reported figures.

As for commercial quality regulation, the information provided should be published and performance across companies should be compared. Companies should therefore also be informed about the form and means of publication of the reported data.

#### 4.1.2.4 Validation

If possible the regulator should arrange audits to be conducted on the information that companies have registered and reported. Auditing the data collected on continuity of supply is extremely important because the financial exposure with respect to the measured indicators can be quite large both for the regulated companies and for customers.

**Table 4.4.** Annul SAIDI values, Urban and Rural areas

| URBAN | Trans. | HV | MV | LV | Total |
|---|---|---|---|---|---|
| FM | 0 | 6.4 | 57.1 | 0 | 63.5 |
| All others | 0 | 0 | 0 | 0 | 0 |
| Total | 0 | 6.4 | 57.1 | 0 | 63.5 |

| RURAL | Trans. | HV | MV | LV | Total |
|---|---|---|---|---|---|
| FM | 0 | 7.5 | 0 | 0 | 7.5 |
| All others | 0 | 0 | 67.5 | 12.5 | 80.0 |
| Total | 0 | 7.5 | 67.5 | 12.5 | 87.5 |

[8] Similarly, the annual ENS will be the sum of the $ENS_i$ over all the interruption events in the year.

**Table 4.5.** Annul SAIFI values, Urban and Rural areas

| URBAN | Trans. | HV | MV | LV | Total |
|---|---|---|---|---|---|
| FM | 0 | 0.43 | 0.57 | 0 | 1.0 |
| All others | 0 | 0 | 0 | 0 | 0 |
| Total | 0 | 0.43 | 0.57 | 0 | 1.0 |

| RURAL | Trans. | HV | MV | LV | Total |
|---|---|---|---|---|---|
| FM | 0 | 0.50 | 0 | 0 | 0.50 |
| All others | 0 | 0 | 1.0 | 0.12 | 1.12 |
| Total | 0 | 0.50 | 1.0 | 0.12 | 1.62 |

To this end, companies should be required to keep all the necessary documentation (above all, the register) that could affect the reporting of performance. Furthermore, the regulator should be informed about the presence of a Supervisory Control And Data Acquisition (SCADA) system on the company's distribution network. These systems enable the distributor to supervise the network (substations and feeders) from one control centre. In addition, SCADA systems can store a large amount of data on events and measures from the network. In other words, they automate the data recording process.

As far as auditing is concerned, the fact that SCADA systems can automatically provide a log of all actions performed on the relevant components of the network is extremely important. Auditors will be able to compare the log and the register and to verify that all interruptions were actually registered and that, for instance, the duration of the event was registered correctly (on audits, see also Chapter 6). When there is no SCADA system in place, regulators must be aware that they are relying only on the accuracy of the regulated companies in keeping track of the interruption events.

While telecontrol and SCADA systems are normally installed on the high voltage portion of the network,[9] they are sometimes also used at medium voltage levels. Often, however, only a portion of the medium voltage network is equipped with such systems.[10] A regulator can introduce obligations or incentives for companies to gradually control and operate the entire medium voltage network with telecontrol and SCADA systems. Nonetheless, accuracy in data collection should not be the main driver for this innovation. Obligations and incentives should be introduced only if the benefits to be gained in terms of network reliability are higher than the costs of the systems.

As SCADA systems are used on larger portions of the medium voltage network, the regulator will benefit from greater accuracy in data collection. Indeed, it is not uncommon to observe a decline in the measured performance of companies right after the adoption of these systems.

[9] SCADA systems provide information on the status of the network components, while telecontrol systems enable the distributor to remotely perform a number of operations on network components (for example, opening and closing circuit-breakers). Obviously, the two systems are strongly coordinated.

[10] By contrast, it is not technically or economically efficient to equip low voltage networks with telecontrol and SCADA systems.

### 4.1.3 Long interruptions: individual level

Measuring the number and duration of long interruptions at the individual level is more complex than measuring system-level indicators. The main difficulty lies in the *identification of customers affected* by the interruption event.

The identification is only possible if the company has an adequate connectivity model which is kept up to date. As illustrated in Figure 4.1, the same customer can be supplied from different substations, depending on the configuration of the distribution network. Most of the time companies operate the network in a *standard configuration* mode. However, the configuration can be modified for different reasons. Thus, in order to register long interruptions at the individual level the distributor needs a connectivity model that includes the *current* configuration of the network.[11]

Another difficulty lies in the high number of customers served by companies. For the latter, keeping a register of individual supply interruptions for very large numbers of customers could be too time-consuming, and thus costly.

Individual measurement of long interruptions should be approached gradually. A gradual approach can be achieved by individual measures applicable only to larger customers (i.e., those connected to the high and medium voltage feeders of the distribution network)[12], and by simplifying reporting rules so that companies can be obliged, for example, to communicate a list of all interruptions to the affected customers (and not to the regulator). The information reported to the regulator can be in aggregate form. For instance, the report could include only the number of end-users who experienced one interruption in a year, the number who experience two interruptions and so forth. This enables the regulator to have a general picture of the overall performance, without entering into the details of individual cases.[13] Secondly, gradualness can be achieved by introducing individual data collection requirements for the larger distribution companies, thus giving smaller network operators, which may have fewer resources to dedicate to the task, more time to adapt to the regulation.

A separate discussion regards the introduction of Guaranteed Standards (GS) with respect to the number or duration of interruptions for the individual customer (see Section 4.3). The data collection requirements for these regulatory instruments would be extremely demanding if companies automatically had to compensate individual customers for any breach of performance standards. By contrast, if

[11] The UK regulator indicates that a reasonable time frame for updating the connectivity model is within 14 days of any permanent changes in the configuration of the network or regarding customer connections. For example, a network change expected to be in place for at least 28 consecutive days may be regarded as a permanent change. In addition, the numbers of customers in the model could be reconciled with the total number of connected customers on a monthly basis [Ofgem, 2005b].

[12] These customers are far less numerous than those connected to lower voltages.

[13] Separate reporting sheets could be requested for high and medium voltage customers and for different geographical areas [AEEG, 2004].

compensation is paid on request, these regulatory instruments can be employed (and are employed in practice) even when individual output measures are not available.

In summary, measuring long interruptions at the individual level should not be a first priority for regulators approaching continuity of supply regulation for the first time. A gradual approach to the issue of data collection is certainly a preferable solution. In addition, the lack of such measurements does not prevent the adoption of regulatory mechanisms in the form of GS.

### 4.1.4 Short interruptions

There are several countries where instructions and guidance are given for the collection of data on short interruptions and where the corresponding indicators are periodically published. However, to date the use of regulatory instruments in this area has been rather limited. Only recently has the problem of the number of short interruptions been prioritized among customers and regulators as well. Industrial customers are particularly sensitive to short interruptions. Indeed, the consequences of a short interruption event for a manufacturing plant can be as serious as those of a long one.

Before introducing regulatory instructions on data collection, it is important to understand the basic technical aspects regarding the events being recorded. This section contains only a brief description of these aspects and we refer the reader to the engineering literature for further information.

According to the EN 50160, accidental interruptions are classified as *short interruptions* when supply is restored in less than three minutes [CENELEC, 1999].

Short interruptions are often caused by a transient fault. Most faults on overhead lines are transient: they require the intervention of the protection system, but they do not cause permanent damage to the system. A frequent cause of a transient fault is a small object causing a temporary path to ground (a transient short circuit). The object, for instance a small branch from a tree, will either drop to the ground or burn up due to the high current during the fault, leaving only an arc. Soon after the protection system removes the faulted line from the network the arc disappears. The automatic reclosing of the circuit-breaker (after a given reclosing interval) can restore the supply without any permanent damage to the system. However, there is a risk that the fault will not clear within the reclosing interval. In that case, the protection system will trip a second time and give the fault a second chance to restore supply by means of a longer reclosing interval [Bollen, 2000]. When this 'multi-shot' reclosing scheme restores supply in less than three minutes, the resulting interruption is classified as 'short'.

Even in the case of a permanent fault in the system, a number of customers may experience only a short interruption. This may occur when the network equipment enables a prompt isolation of the fault and a restoration of supply to some of the affected customers by recourse to of an alternative, healthy supply ('back-feeding').

Hence, in the same way as for long interruptions, a short interruption is the consequence of a fault clearing action by the protection system. Whether the restoration is achieved automatically by reclosing the circuit-breaker or is performed by a remote control operation of the switchgear, in less than three minutes, the resulting event is a short interruption. The fact that short interruptions result from automatic reclosing or remote control operations means that recording them requires some kind of monitoring equipment to be installed on the feeders (otherwise the events will go unrecorded).

Indeed, the first difficulty in collecting data on short interruptions lies in counting the *number of interruptions*. A distinction should be made here between networks operated with a SCADA system and networks that are not.

When a SCADA system is in place, the distribution company will register, as the start time of the interruption, the instant at which the switchgear on the distribution system removes the faulted line from the system (opening the circuit-breaker). Registering the end time of the interruption is not always required, as long as supply is restored in less than three minutes. For instance, in the case of multi-shot reclosing schemes (which terminate in less than three minutes), only one short interruption will be counted even if successful restoration is achieved by a sequence of multiple operations.[14]

When a SCADA system is not in place, companies can use counter readings on reclosing devices to register the number of interruptions. When this is the case the counters should be read with a given periodicity (e.g. annually) and companies will not be required to register the dates and times of short interruptions [Ofgem, 2005b].

However, even when a SCADA system is in place, counting the number of interruptions can be problematic. Short interruptions result from an operation of the protection equipment of the type described above; thus, they often come in *sequences*, and/or precede or follow longer interruption events. Regulators need to define instructions on the treatment of these sequences of events. On the one hand, to date, these instructions are quite different across countries. On the other hand, these instructions can significantly influence the outcomes of the process of data collection. As an example, we summarize below the instructions for three different countries: the UK, Italy and France [CEER, 2005].

### United Kingdom

Short interruptions that precede long interruptions must be recorded as separate interruptions. Short interruptions that follow a long interruption must be recorded as separate interruptions only if they occur more than 3 hours after the end of the long interruption.

Short interruptions occurring for some customers during a long interruption for other customers on the same circuit must not be recorded.

### Italy

Short interruptions that follow a short or long interruption must be recorded as separate interruptions only if they occur more than 3 minutes after the end of the previous interruption.

[14] As for long interruptions, the company can be required to register their origin and cause.

Short interruptions occurring for some customers during a long interruption for other customers on the same circuit must be recorded as separate interruptions.

**France**

Short interruptions that precede a short or long interruption must be recorded as separate interruptions only if they occur more than 2 minutes before the beginning of the following interruption.

Short interruptions that follow a long interruption must be recorded as separate interruptions only if they occur more than 1 hour from the beginning of the previous interruption.

It is evident that, even when simply reading a report on short interruptions, one should always bear mind the rules for interruption sequences. In particular, comparative reports of data from different countries should be interpreted carefully.

A second reason for complexity in collecting data on short interruptions, is the *identification of the customers affected*. Estimates are often used, based on the assumption that the circuit was configured normally.

As for long interruptions, when the exact or estimated number of customers affected is available, we can compute *system-level indicators* of performance. The 'average number of short interruptions per customer per year' is calculated in the same manner as the SAIFI indicator, where long interruptions are substituted with short ones. This indicator is internationally known as MAIFI (Momentary Average Interruption Frequency Index):

$$MAIFI = \frac{\sum_{i=1}^{K} N_i}{N_{tot}}$$

Note that, an equivalent of the SAIDI indicator is not computed for short interruptions: their actual duration is not particularly relevant for customers.

When a connectivity model is in place, we can also calculate *individual indicators* of performance. The feasibility of such calculations will depend on the level of accuracy of the connectivity model. As for long interruptions, it is advisable to begin with individual measurements of high and medium voltage customers and to allow companies, at least for a transitory period, to assume that the network is always operated in a standard configuration mode.

Finally, individual indicators of performance should be *reported* to the affected customers and system-level indicators should be reported to the regulator.

In summary, even regulators who approach continuity of supply regulation for the first time should consider introducing instructions for measuring short interruptions. Industrial customers can be very sensitive to these events. When introducing such an obligation, it is important to analyze the characteristics of the monitoring equipment on the network (for instance, it is important to know which portion of the network is equipped with a SCADA system), and the precision of the connectivity models used by the distribution companies. As elsewhere, a gradual approach to the issue is, in all cases, the preferable solution.

## 4.2 Publication

Comparative publication of performance data reported by companies does not require the regulator to set performance standards or financial incentives. Nonetheless, comparative publication has a strong impact on the reputation of the regulated companies and it has proved an extremely effective regulatory instrument. Most regulators publish annual statistical indicators on long interruptions at system-level. Aggregate figures can also be given on long interruptions measured at the individual level as well as on the number of short interruptions, when available.

The methodology used in the collection of continuity of supply data can significantly influence the values of the indicators provided by companies. For this reason, publication of performance data, and in particular of comparative performances, is only significant when the data are collected in a uniform manner by all companies. Therefore, publication should be limited to those figures measured according to the regulatory instructions and guidance or should be deferred until the instructions have been adopted by the regulated companies. Alternatively, differences in the measuring protocols used by different companies should be indicated.

Data communicated to the public can relate exclusively to the measured quality indicators, in the form of a comparison of performance across different companies or as an analysis of performance over time. Additional interesting information can be given once other regulatory instruments are in place (MQS, or reward and penalty schemes). In particular, the measured indicators should be compared to the performance standards defined by the regulator. Furthermore, the financial costs of the regulatory instruments (compensation payments and penalties paid by companies and rewards paid by customers) should also be part of the communication.

In practice there are several ways to communicate continuity of supply data to the interested parties. One is to include such data in the regulatory authority's annual report which usually highlights the national trend in performance over time. A second way is to prepare annual specific reports on the performance of distribution companies and to make them available for downloading on the regulatory authority's website. A third means is to oblige distribution companies to communicate their performance directly to customers in a note attached to the electricity bill once a year. Fourthly, these figures can be made available on the web site of the regulatory authority, in the form of a database. This allows all interested parties to access the data using simple queries (for instance, regarding a particular company or year).

## 4.3 Minimum quality standards

Minimum quality standards (MQS), in particular in the form of Guaranteed Standards (GS), are regulatory instruments employed to ensure that a minimum level of quality is met in each individual case.

Before describing how these instruments are designed in practice, it is worth noting that both individual and system-level indicators are measured when evaluating continuity of supply. In turn, only one commercial quality indicator is normally measured for each of the regulated services: either an individual one (in most cases the time necessary to provide the requested service), or an average one (the quality of call centres). Continuity of supply is highly valued by customers: controlling both individual and system-level indicators is important because they play different roles. As observed in Chapter 3, measured individual quality indicators are distributed across customers with a probability distribution. Regulating system-level indicators is equivalent to controlling the average value of the distribution. Improvements obtained on the average value are highly desirable because they increase the expected level of quality for all customers. However, even a good average performance does not necessarily prevent a portion of the customer base from receiving unacceptable levels of service quality [Rivier and Gomez, 2003]. The number of worst-served customers depends on the variance of the distribution and not on the average (see Figure 4.4). GS which define performance standards to be met in each individual case, ensure that the variance is kept small. By contrast, reward and penalty schemes which define performance standards on system-level indicators, focus on the level of quality provided on average across the customer base.

In continuity of supply regulation MQS generally take the form of GS. When designing these regulatory instruments, regulators are required to make a number of decisions regarding:

- *individual quality indicators that are subject to regulation.* In commercial quality regulation the corresponding decision was made in relation to the services to be regulated. Here we are dealing with a single service. However, the quality of the service is captured by two dimensions: the number of events and the duration of supply interruptions. In addition, interruptions are of different types and

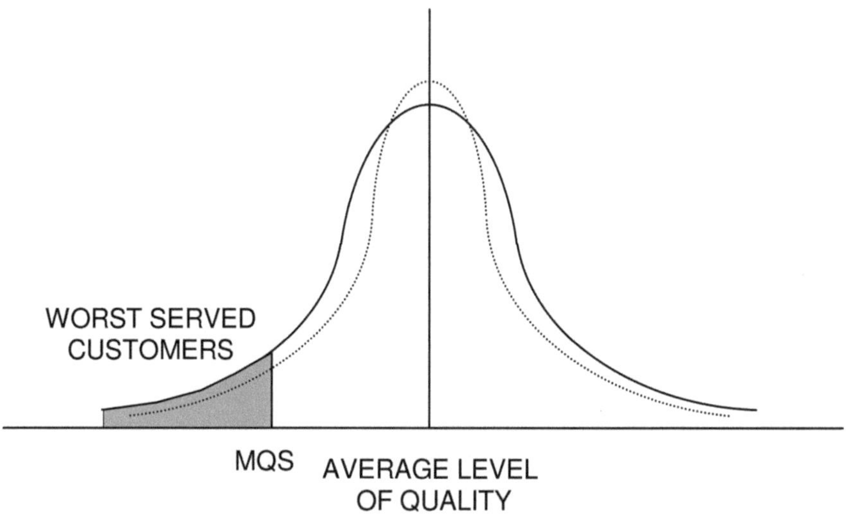


**Fig. 4.4.** Distribution of quality levels

durations. Accordingly, several indicators can be regulated at the individual level;
- *performance standards* to be met in each individual case and any exemptions;
- *compensation payments* due to customers who do not receive the required level of service quality.

An additional question posed by the design of GS is the availability of measured, individual quality indicators. In Section 4.1 we observed that distribution companies can have difficulties in measuring performance at the individual level. Nonetheless, we also mentioned that GS can be introduced when customers make their own claims for compensation. In this case, customers 'measure' the quality that they receive and 'supervise' compliance with the performance standards set by the regulator.

The above issues are discussed further in Section 4.3.1, which deals with individual quality indicators, and in Section 4.3.2, which examines performance standards and compensation payments.

### 4.3.1 Regulated indicators

Several European countries have introduced GS. Observation of the international experience gives rise to the following considerations.

The individual indicators that are subject to regulation can be classified into two groups: one focusing on the interruption duration; and the other focusing on the number of interruptions.

Concerning the first group, CEER (2005) reports that the regulated indicators are:

- duration of a single long, unplanned interruption;
- cumulative annual duration of long unplanned interruptions.

Customers are highly sensitive, in particular, to the duration of a single, unplanned interruption. For this reason, numerous regulators apply GS to this quality dimension, although each country has its own conditions for enforcement. By contrast, the second indicator is less frequently employed.[15]

Concerning the second group, CEER (2005) reports that the regulated indicators are:

- annual number of long unplanned interruptions;
- annual number of short interruptions;
- annual number of interruptions (short and long).

The number of interruptions experienced in a year is another aspect of primary importance for customers. The difficulties already mentioned with this type of measurements explain why GS in this group are less popular and generally relate only to long interruptions and/or customers connected to higher voltage levels.

[15] There are also GS for planned interruptions although they are not often used.

### 4.3.2 Performance standards and compensation payments

Although there is already considerable experience with GS, it tends to be extremely varied. The CEER *Benchmarking Report* (2005) provides detailed information on the GS in use. We focus here on two of the most frequently employed: the GS on the 'duration of a single unplanned interruption' and the GS on the 'number of long unplanned interruptions'.

As regards the *duration of a single unplanned interruption* the choices made by regulators in terms of performance standards are quite different, ranging from 4 hours in Belgium to 24 hours in Lithuania and Estonia. Moreover, performance standards can differ between summer and winter periods (in Estonia, respectively 20 and 24 hours), between single and multiple disturbances (in Hungary, 12 hours and 18 hours respectively), and between voltage levels (in the Czech Republic, 18 hours for lower voltage and 12 for higher voltage). The protection granted to final customers by these instruments should be analyzed taking into consideration the combination of performance standards, the exemptions, and the amount of compensation payments.

In all cases reported by CEER (2005) exemptions apply for exceptional events (*force majeure*). This is consistent with the idea that some events are beyond the control of the distribution company. However, the disturbance created by long unavailability of supply, particularly in cases of severe weather conditions, can be extremely high for end users. For this reason, regulators have recently reconsidered granting exemptions in these cases. The idea is that a distinction can be made between the occurrence of the event (a fact that is beyond company control) and the actions taken to promptly restore service (a task that is largely under the company's control). Consequently, a regulatory mechanism is applied in UK (and is being considered in other countries) to induce companies to shorten the restoration time in cases of severe weather conditions. The interested reader can consult the actual decision of the regulator [Ofgem, 2004a].

Compensation payments are generally made on request. Automatic compensation would provide stronger incentives for the regulated utility to comply with the standards; however, when individual indicators are not available, compensation on request is the only possible choice. In this case, the regulator can introduce other mechanisms to encourage companies to be pro-active in making payments. In the UK, a company failing to meet a performance standard is always penalized. The company either pays compensation to customers (on receiving a request) or, when customers do not claim compensation, it faces an equivalent reduction in its price control revenue.

Compensation generally increases as a function of the 'distance' from the performance standard and it is differentiated on the basis of customer types (domestic/non-domestic). This is consistent with the idea that the damage is dependent on contractual capacity. For instance, in Sweden a domestic customer (and small industrial customer) receives 12.5% of the annual network charges for interruptions in the 12–24 hour range, 37.5% for interruptions lasting 24–48 hours, 62.5% for interruptions lasting 48–72 hours, and an additional 25% for every 24 extra hours (the minimum compensation is ~84 euros/interrup. and a cap is set at ~315 euros/interrup.). Compensation scaled on the annual network charge automatically differentiates payments on the basis of customer type.

No specific rules apply to the selection of the performance standards and compensations. In any case, the decision on the performance standard should be calibrated with the decision on the monetary reimbursement. For instance, the French standard is set at 6 hours (with exemptions for exceptional events), and the reimbursement amounts to a few euros for a domestic customer. The UK standard is set at 18 hours, but amounts to 36 euros for the same type of customer (severe weather conditions in UK are treated separately from normal weather conditions but do not lead to exemptions). Customer surveys are a good source of information regarding the costs of very long interruptions. An *ex ante* assessment of regulation can provide insights into the most efficient design of GS for each country.[16]

Regarding the *number of long unplanned interruptions*, performance standards on multiple interruptions vary significantly across European countries, ranging from 2 (in France) to 8 (in Estonia) long interruptions per year for medium voltage customers in urban areas, and from 5 (in Italy) to 25 (in Portugal) for medium voltage customers in rural ones. Different performance standards apply for the other voltage levels, but the variety remains large in all cases. Several exemptions apply for all mechanisms reported by CEER (2005), particularly for exceptional events.

The same recommendations made above apply to the choice of the performance standards and compensations. Two additional aspects are particularly important. The first is the treatment of sequences of interruptions. These can create ambiguities in the calculation of the number of events. Secondly, in setting the performance standards it is important to account for differences in the expected performance according to geographical areas or network design (urban/rural areas, overhead lines/cables).When individual measures are available, a possible approach for setting multiple interruption performance standards is to look at interruption statistics and to select the number of events above a given percentile. For instance, Italian data indicated that over 90% of MV urban customers experienced less than 3 long interruptions per year and above 90% of rural ones experienced less than 5.[17] Therefore, the performance standards for the regulatory period were set at 3 and 5 respectively (performance standards are reviewed every four years) [AEEG, 2004].

Compensation is paid both automatically and on request, and it can take different forms. This generally depends on the 'distance' from the performance standard (sometimes subject to a cap), and further depends on the capacity contracted by

[16] An *ex ante* assessment of the regulation is an analysis of the impact of the regulatory decision on the interested parties. In this case it would be an estimate of the expenditure that companies will sustain as a consequence of the decision regarding the amount of an individual compensation and the expected number of payments due (based, for instance, on company performance over time). See also Section 6.1.

[17] Exemptions lead to the exclusion of few interruptions.

the customers. However, it can also be in the form of a fixed amount (differentiated according to classes of customers). A compensation structure of the first type is employed, for instance, in Spain:

$$\text{Compensation}_{e,c} = PW \cdot H \cdot P_{e,c} \cdot \frac{\Delta N}{8}$$

where:

$PW$ contractual power in kW
$H$ cumulative duration of the interruptions in hours (for the interruptions in excess of the performance standard)
$P_{e,c}$ annual average price for respectively eligible, $e$, and captive, $c$, customers, in euros/kWh
$\Delta N$ difference between the actual number of interruptions and the performance standard.

In summary, GS applying to the maximum duration of a single unplanned interruption provide incentives for the regulated distribution company to limit the consequences of very long interruption events. The event may be unpredictable and unavoidable; however, a prompt restoration of supply can sometimes be achieved if adequate means are devoted to the problem. GS concerning multiple interruptions provide strong, location-specific signals for the regulated distribution company to invest in network quality enhancements. Even if only applied to MV customers, this instrument can also generate benefits for LV customers, insofar as most faults leading to customer interruptions originate at the MV level of the distribution network.

## 4.4 Reward and penalty schemes

Reward and penalty schemes were briefly discussed in Chapter 3. They were presented as complex instruments that regulators apply to system-level indicators in order to induce the regulated company to deliver desirable levels of service quality. Incentive schemes modify the company revenues according to its performance against performance standards set by the regulator. In doing so, they tend to replicate the outcome of a competitive market, associating higher quality levels with higher revenues and vice versa.

In this section we introduce reward and penalty schemes more specifically, beginning with the theoretical reference that regulators have in mind when they design them.[18] We then turn to a more practical view clarifying what can be done concretely with such mechanisms (Section 4.4.1). With the help of available experience

[18] We chose to include this theoretical view in the chapter concerning continuity of supply because in Europe reward and penalty schemes generally apply to this area of service quality. However, this theoretical introduction has general validity and is relevant for any area of service quality.

we describe the numerous practical issues involved in the design of an incentive scheme of this sort. Each step corresponds to a regulatory decision regarding respectively:

- definition of regulatory *objectives* (Section 4.4.2);
- choice of *quality indicators to be regulated* (Section 4.4.3);
- definition of annual performance standards over a period of several years (or *baseline*) (Section 4.4.4);
- characterization of the *functional relation between quality and revenues* (Section 4.4.5).

Existing incentive schemes focus on system-level indicators of performance, calculated on the basis of long interruptions. As will be clear from the following discussion, a regulator is not in a position to introduce a regulatory instrument of this type unless he can count on at least a couple of years of reliable data corresponding to the relevant indicators.

### 4.4.1 The theoretical reference and the scheme in practice

Providing quality has a cost for the regulated utility, given by both investment and operation and maintenance expenditures. This cost increases for higher levels of quality and, as indicated in Figure 4.5, unitary improvements in quality are more costly to achieve when the level of quality is already high. In the same way, customers benefit from increasing quality levels: customers benefit a great deal from unitary increments in quality when quality is poor, but the benefits gained from unitary improvements in quality gradually decrease at higher levels of quality. Those benefits that consumers receive from quality represent their willingness to pay (WTP) for it. Hence, from a social perspective (which includes the interests of both

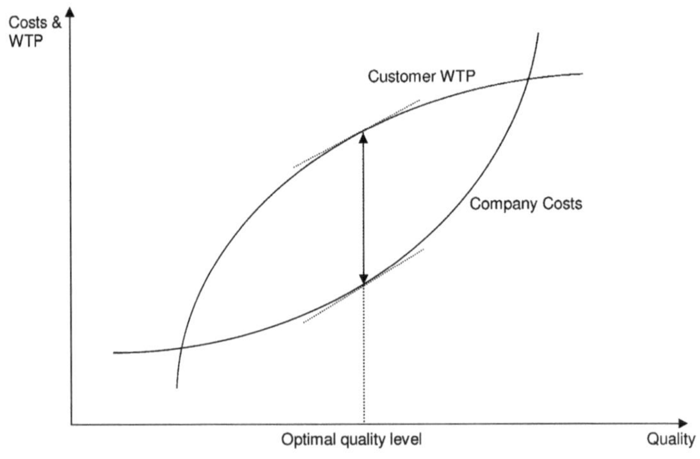


**Fig. 4.5.** Optimal quality level, I

consumers and companies), the *optimal level of service quality* is the level at which the marginal benefit of additional quality equals the marginal cost of supplying it. As illustrated in Figure 4.5, this is the level of quality where the tangent to the consumer WTP curve and the tangent to the company cost curve are parallel.

Quantifying WTP for quality is rather difficult. In practice, WTP is usually approximated by its inverse: the costs of the customer incurred as the result of a poor level of quality. Another way of expressing the same concept is thus illustrated in Figure 4.6: the optimal level of service quality corresponds to the minimum of a total cost function, that is, the sum of the company's costs incurred in providing quality and the costs incurred by customers due to poor quality.

The objective of a regulator is to achieve this optimal level of quality. However, in order to do so, a regulator needs to know the cost curves of both the customer and the company. Customer surveys are a major source of information regarding the so-called 'direct costs' incurred by customers.[19] Engineering studies can be a good reference for the costs of the company in providing service quality. However, for obvious reasons, the regulated company generally has better information than the regulator about the exact costs of supplying quality in practice [Sappington, 2005].

As a matter of fact, an incentive scheme of the type discussed here enables the regulator to achieve the optimal level of quality by making good use of the company's favourable position. In order to do so, a financial incentive scheme specifies (a performance standard and) rewards and penalties that reflect customer valuations of quality. "If the bonuses and penalties presented to the firm closely approximate the marginal benefits and costs to consumers of increases and decreases in quality, a profit-maximizing regulated firm will expand quality to the point where the marginal benefit of additional quality to consumers (and thus

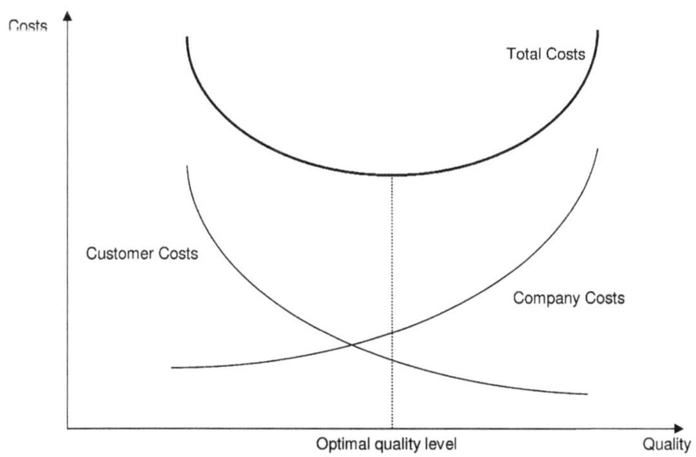


**Fig. 4.6.** Optimal quality level, II

[19] On customer surveys, see Chapter 6.

the firms marginal reward) equals the firms marginal cost of increasing quality" [Sappington, 2005: 134].

In other words, the reward and penalty scheme forces the regulated company to internalize customer costs in its own cost function. The company will make decisions based on the total cost curve in Figure 4.6, and not only on the basis of the costs the company itself incurs for providing quality. As a result, the company will deliver an efficient level of service quality from a social perspective. For this reason, a good estimation of customer valuation of quality is a key element of reward and penalty mechanisms.

The discussion on incentive schemes has been rather theoretical insofar as we assumed both quality and the valuation of it by customers to be mono-dimensional. In practice, quality is multi-dimensional, and customer valuations of quality are complex functions of the different quality dimensions. Thus, in practical applications, reward and penalty schemes will not induce the regulated company to deliver precisely an optimal level of service quality. Reward and penalty schemes in practice will focus on and reflect customer valuations of a small number of quality dimensions and, for these few quality dimensions, they will induce the regulated company to deliver,

- a higher level of quality than specified in the performance standards set by the regulator when the benefits which customers derive from increased quality (for the company: financial rewards) outweigh the associated costs to the company and, conversely,
- a lower level of quality than specified in the performance standards set by the regulator, when the cost savings outweigh the associated losses incurred by customers (for the company: financial penalties) [Sappington, 2005].

In summary, when designing a reward and penalty scheme, a regulator must bear in mind that, in practice, the regulatory instrument will act on a small number of quality dimensions, and that the company will deliver a level of quality that depends on the choices made by the regulator concerning the level of financial incentives and performance standards. Hence, the choice of quality indicators to be regulated, the respective performance standards, and the amount of financial penalties and rewards will be key elements in the overall design of applied mechanisms. The design of these schemes in practice will thus be the focus of the rest of this section.

### 4.4.2 Objectives

Reward and penalty schemes respond to the general objective of ensuring that desirable levels of service quality are delivered to customers. The adoption of this regulatory instrument is often motivated by the need to counteract the potential risk of quality degradation related to the adoption of price cap regulation, on the one hand, and privatization of the distribution companies on the other. In addition to this general concern, when introducing a reward and penalty scheme regulators often specify one or more precise objectives to be achieved with this regulatory instrument.

We report on a few of the more common ones below [CEER, 2005].

Depending on the circumstances, the current level of continuity of supply provided by distribution companies may be perceived as unsatisfactory in relation to the performance of other distribution companies inside or outside the country. In this case, regulators design incentive schemes that promote increases in the average level of service quality (this is done with an appropriate choice of performance standards).

An assessment of the initial levels of quality can highlight specific problems. For instance, in Italy, a large difference in performance was observed between the North and the South of the country. Hence, the incentive scheme was designed to promote higher improvement rates in the South than in the North.[20] In other countries too, the incentive scheme is specifically designed to promote a convergence of performance levels across different distribution areas.

Quality regulation in the form of incentive schemes focuses on the average level of quality. Regulators are also concerned with protecting the worst-served customers. As explained in Section 4.3, this can be achieved using GS. As an alternative, regulators sometimes introduce special provisions targeting the worst performing areas (instead of worst-served individual customers). In Sweden, the regulatory intervention consists of a close monitoring of the performance of the worst-served areas. In Spain and Portugal, companies serving these areas are given the opportunity to present special quality-improvement plans which, if approved by the competent authority, are partly financed through tariffs. If the company does not meet the objectives indicated in the plan or if the plan is delayed, the company can be penalized [Rivier and Gomez, 2003].

Incentive schemes mimic a competitive market by assigning higher revenues for higher continuity levels. It is interesting to note that, in the case of special plans, this logic is reversed, as financial support is given to the worst-performing areas for investment in network improvement. This choice is based on a regulatory judgement regarding the company's financial resources: if these are deemed insufficient to make the necessary investments, then an aid-based scheme is preferable to a reward and penalty mechanism.

### 4.4.3 Regulated indicators

The decision on the quality indicators that are subject to financial incentives is crucial for two reasons. First, offering financial incentives for some indicators of quality and not others may introduce an incentive to neglect those not covered. In addition, objective and verifiable measures must be collected and assessed for the selected indicators [Williamson, 2001].

Continuity of supply is captured by both duration and frequency of interruptions, and interruptions are of different durations and types (see Section 4.1). Several quality indicators can thus be derived from the collected data. However, workable incentive schemes focus on a limited number of them. Regulating

[20] Performance was compared across areas with similar population density: for instance, Southern urban areas and Northern urban areas.

one or two indicators, three at the most, is highly recommended at the beginning. Penalties and rewards can be introduced for other indicators at a later stage.

Experience suggests that there are no more than two or three indicators of continuity of supply that are subject to regulation ('regulated indicators'). In addition, the existing incentive schemes focus on indicators derived from the recording of long interruption events, as these seem to have the highest correlation with domestic customer satisfaction. In several countries the incentives apply to the SAIDI indicator (the average interruption duration per customer in minutes per year). Alternatively, the chosen indicator is the Energy-Not-Supplied (ENS). In other countries, the schemes apply to both SAIDI and SAIFI.

Several practical indications can be derived from an analysis of the applied incentive schemes.

First, a regulator should decide to reward and penalize performance either in terms of unplanned interruptions, or in terms of both planned and unplanned interruptions. In making this decision, one should consider that, on the one hand, a scheme that allows companies to gain higher revenues by reducing planned interruptions can induce companies to adopt a more efficient maintenance programme (and in particular, to plan maintenance when consumption is low). On the other hand, it may also create a long-term risk due to insufficient network maintenance. When planned interruptions are subject to financial incentives, it is customary to give them a different weight with respect to unplanned ones. For example, in the UK a customer survey indicated that their impact on a customer is about half the impact of unplanned ones. Therefore, they are included in the incentive scheme with a 0.5 discount factor. In Norway and Sweden their reduced impact on customers is taken into account using a lower financial incentive rate [CEER, 2005].

Secondly, the regulator should decide what exemptions apply. For instance, interruption events that originated on the transmission network are often subject to exemptions on the ground that the transmission network is operated by a different company. Similarly, the financial incentive scheme does not usually apply to exceptional events. Some regulators have recently begun to adopt other methodologies for exemptions based on the statistical observation of daily continuity data ('statistical methodologies'). Additional information on this interesting, but rather advanced topic, is given in Section 4.6.

Finally, the regulator should be able to identify the measured values of the regulated indicator (i.e., the actual performance) from companies' reports. In particular, as noted in Section 4.1, the report to the regulatory authority should already give evidence of the allowed exemptions. Referring once again to the example introduced in Section 4.1, let us assume the regulated indicator has been identified with SAIDI for long, unplanned interruptions, net of exemptions for events originating on the transmission network and caused by *force majeure*. The measured value of the regulated indicator (e.g., referring to a rural area) can be derived by excluding from the report in Table 4.6 (taken from Table 4.4, Rural areas) the second column (events originating on the transmission network) and the second row (events caused by *force majeure*). Hence, penalties and rewards will be calculated comparing a rural-SAIDI of 80 minutes to the rural-SAIDI performance standard set by the regulator.

**Table 4.6.** Annul values SAIDI, Rural areas

| RURAL | Trans. | HV | MV | LV | Total |
|---|---|---|---|---|---|
| FM | 0 | 7.5 | 0 | 0 | 7.5 |
| All others | 0 | **0** | **67.5** | **12.5** | **80.0** |
| Total | 0 | 7.5 | 67.5 | 12.5 | 87.5 |

### 4.4.4 Baseline

Defining the performance standards of an incentive scheme is a delicate task. Incentive schemes induce the company to deliver, on the regulated indicators, a higher level of quality than specified in performance standards when the rewards that derive from increased quality outweigh the associated costs to the company. Conversely, the company will deliver a lower level quality than specified in the performance standards when the cost savings from doing so outweigh the associated penalties [Sappington, 2005].

Regulators often define a series of annual performance standards in advance for the entire period of validity of the incentive scheme. This series of performance standards is also called the 'baseline'. Defining the baseline requires a considerable amount of work. Incentives schemes should only remunerate (or penalize) variations in quality from the level that a company is expected to deliver, given the remuneration received through the tariff. In other words, the tariff should cover the cost of delivering the same level of quality defined with the baseline. Rewards should only remunerate the costs of increasing quality above the baseline and penalties should reduce the tariff if the quality level falls below it. Therefore, in order to avoid inefficiencies (double remunerations for costs incurred for quality or under-remuneration for costs incurred for quality), the baseline must reflect the expected performance of the company over the period of validity of the incentive scheme, given the level of the tariff. For this reason, it is advisable to define the baseline (the expected performance of a regulated company) at the same time as the price control review and over a similar time span.

When setting performance standards, the regulator should take into account several factors that influence a company's expected performance. They can be grouped into three classes [CEER, 2005]:

- Inherent factors such as weather conditions, geography and population density of a particular area;
- Inherited factors such as the network design at the moment the regulatory instrument takes effect (some companies or areas may have long, predominantly overhead circuits, whilst others may have more underground lines). It takes a long time and significant capital expenditure to alter network design;
- Incurred factors such as managerial performance, maintenance of assets, and effective use of resources.

In practice, there are two ways to set the baseline while taking into account the above factors. The first is to look at company performance over time, as described below. The second is to use a mathematical or engineering model. For instance, a regression model is used in Norway to calculate the performance standard (or the expected ENS) for each company.[21] However, the topic of modelling the network performance for regulatory purposes is beyond the scope of this handbook.

As far as the first approach is concerned, a baseline clearly needs to be calibrated with the company's performance over time in order to be feasible. Expected performance does not need to coincide with the trend of improvement over time (it can be more demanding), but it should take the initial level of quality as a reference in indicating a feasible evolution of quality levels in the following years. The expectation for improvements in the short-term is strongly influenced by incurred factors; however, companies that begin with lower levels of quality can be expected to achieve, within the same time span, greater improvements than companies that already offer a high level of quality. Similarly, companies that begin with lower levels of quality can be expected to achieve, with the same level of expenditures, greater improvements than companies that offer already a high level of quality (see Figure 4.5). In addition, using longitudinal data enables the regulator to account for inherent as well as inherited factors. The expected short-term performance will thus be differentiated according to geography and network design. For all these reasons, incentive systems in practice comprise not one, but a number (sometimes quite large) of different baselines, individually set for each regulated company (or for each geographical area composing a distribution territory).

It is important to note that some of the factors that influence continuity of supply are stochastic, particularly weather conditions. These factors can change significantly from one year to the other. For this reason, company performance must be observed over a time period that exceeds a year: two to three years of data can begin to provide a trend in performance that discounts the influence of stochastic events.

This approach in setting the baseline can be clarified using two examples taken, respectively, from the Hungarian and the Italian regulation.

In Hungary, the baseline requires distribution companies to achieve different annual improvements in SAIDI, depending on the initial level of continuity. Table 4.7 shows that a smaller (1%) annual improvement is expected from the best performing companies (SAIDI: 61–180 minutes) and a larger (5%) annual improvement is expected from the worst performing ones (SAIDI: >360 minutes).

The baseline for Italian distribution companies has a similar structure (it depends on the initial level of performance), but it is more articulated. A better performance is expected in urban areas than in suburban and rural areas. The national territory is in fact divided into territorial districts differentiated according

[21] The Swedish regulator uses a Network Performance Assessment Model (NPAM) for setting network tariffs. This model includes both quality indicators and customer interruption costs. For a description and a discussion of this model see, respectively, Gammelgard and Larsson (2003) and Nilsson (2005).

**Table 4.7.** Baseline, Hungary

| Initial SAIDI [min] | 61-180 | 181-240 | 241-300 | 301-360 | > 360 |
|---|---|---|---|---|---|
| Annual improvement | 1% | 2% | 3% | 4% | 5% |

to population density (a district is an administrative area that is smaller than the distribution territory of a company and is homogeneous in population density). The improvement rates for the 2000-2003 price control period are given in Table 4.8. For instance, the annual performance standards of an urban area with an initial SAIDI of 120 minutes would be:

- in 2000: 120-10 % = 108 minutes
- in 2001: 108-10 % = 97 minutes
- in 2002: 97-10 % = 87 minutes
- in 2003: 87-8 % = 80 minutes.

Finally, it is important to note that incentive mechanisms have a dynamic nature and that the baseline needs periodic adjustments. In particular, the baseline should be rescaled at the beginning of a new price control period, to the actual level of quality attained by the company (see Figure 4.7 for the Italian reward and penalty scheme). With this adjustment the regulator transfers, to the customers, the benefits of the quality improvements realized over time. This dynamic is not different from the periodic adjustment of the price cap [Ajodhia and Hakvoort, 2005].

### 4.4.5 The relationship between quality and revenues

An incentive scheme must specify a functional relationship between the revenues and the performance of the company. These structures generally require the definition of an 'incentive rate'. Below we discuss, first of all, several issues related to the practical use of incentive rates. Then, we observe how the relationship between revenues and quality can take a variety of different forms. In doing so, we provide details on the incentive schemes applied in different countries.

**Table 4.8.** Baseline (2000-2003), Italy

| Initial SAIDI [min] | | | Annual improvement |
|---|---|---|---|
| Urban areas | Sub-urban areas | Rural areas | |
| ≤30 | ≤45 | ≤60 | 0% |
| 31-60 | 46-90 | 61-120 | 5% |
| 61-90 | 91-135 | 121-180 | 8% |
| 91-120 | 136-180 | 181-240 | 10% |
| 121-150 | 181-270 | 241-360 | 13% |
| > 150 | > 270 | > 360 | 16% |

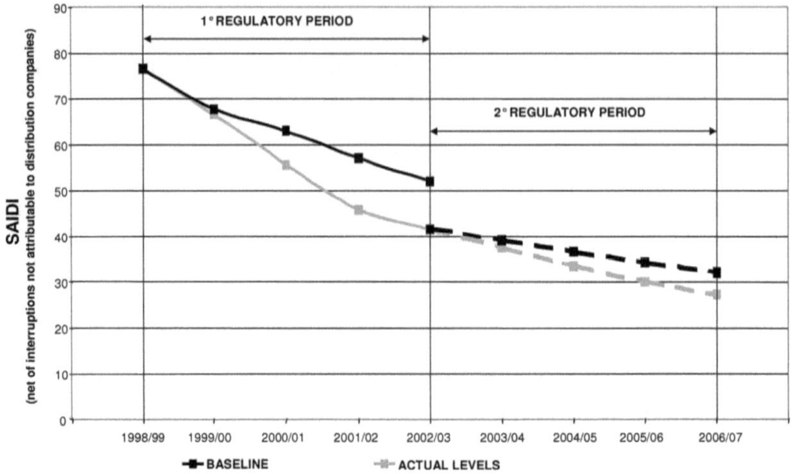


**Fig. 4.7.** Baseline adjustment, Italy

Figure 4.8 illustrates the basic functional relation between revenues and quality: a continuous, linear function. Under this structure, a different financial incentive (reward or penalty) corresponds to each level of quality provided by the company, and it is calculated in the following manner:

- an incentive rate is defined: a monetary value per unit change in quality (e.g., in euros/ENS);
- the 'distance' between the measured level of quality and the performance standard (on the quality axis, e.g., in ENS) is multiplied by the incentive rate.

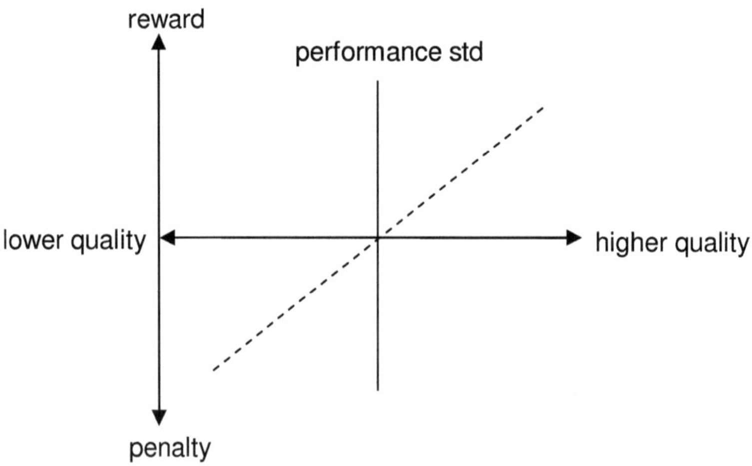


**Fig. 4.8.** Linear incentive scheme

The use of an incentive rate demands attention on two counts: the definition of the incentive rate and the choice of regulated indicator.

As far as the first aspect is concerned, and as explained in Section 4.4.1, for the regulatory instrument to be effective, the incentive rate should reflect customer valuations of higher and lower levels of quality. These valuations are best obtained through customer surveys.[22] Different customer categories incur different interruption costs and have different WTP for quality increases. Moreover, notified and un-notified interruptions cause different levels of disturbance to customers. Thus, results of customer surveys are not easy to translate into a workable number of incentive rates.

Nonetheless, customer surveys remain the preferable method to define the value for customers of the 'energy-not-supplied' and they have been used in several countries (the UK, Italy, The Netherlands, Norway and Sweden). As illustrated inTable 4.9, these values are usually given in [euros/kWh-not-supplied] for a small number of customer and interruption categories.

Note that normalization effects (the weighting of the interruption costs, or WTP, over the energy consumption) can introduce significant differences when equivalent customer categories present different average levels of consumption in different countries. Consequently, an international comparison of incentive rates is always quite difficult to make. The figures given in Table 4.9 for domestic customers range from 1.5 euros/kWh in Portugal to 12 euros/kWh in Sweden (for urban customers and un-notified interruptions).

In practice, regulators have managed to define incentive rates even with extremely limited knowledge of customer costs/WTP. Regulators have often relied on the idea that, as long as quality is thought to be below the efficient level, penalties

**Table 4.9.** Incentive rates

| Country | Value of energy-not-supplied |
|---|---|
| **Italy** | Domestic: 10.8 Euro/kWh |
| *(Un-notified)* | Business: 21.6 Euro/kWh |
| **Norway*** | Industrial: 8.25 - 5.75 Euro/kWh |
| *(Un-notified- Notified)* | Trade/Service: 12.38 - 8.50 Euro/kWh |
| | Agricultural: 1.88 -1.25 Euro/kWh |
| | Residential: 1.00 - 0.88 Euro/kWh |
| | Public service: 1.63 - 1.25 Euro/kWh |
| | Energy intensive industry: 1.63 - 1.38 Euro/kWh |
| **Portugal** | All consumers: 1.5 Euro/kWh |
| *(Un-notified)* | |
| **Sweden** | Urban: 12 - 8.6 Euro/kWh |
| *(Un-notified- Notified)* | Suburban: 8.8 - 6.3 Euro/kWh |
| | Rural: 7.4 - 5.2 Euro/kWh |

* Exchange rate: 1 Euro= 8 NOK

[22] On the estimation of these values through customer surveys see Chapter 6.

and rewards should be large enough to provide meaningful incentives for companies to achieve an increment in the quality they deliver [Williamson, 2001]. In other words, regulators have set incentive rates that were at least greater than the costs incurred by a company to provide a unitary improvement in quality. This information can be obtained from the companies themselves or from engineering studies, without conducting a customer survey.

As for the second aspect, i.e., choice of the regulated indicator, a few considerations are necessary to emphasize the difference between using ENS rather than SAIDI in the incentive scheme. The SAIDI indicator is derived from the duration of the interruption and the number of customers affected by it. In turn, the ENS indicator is calculated using load curves. This difference becomes important in the actual calculation of penalties and rewards.

When the regulated indicator is the ENS, financial incentives (in euros) are calculated multiplying

- the measured ENS (expressed in [kWh]) by
- the value of energy-not-supplied in [euros/kWh-not-supplied].

In turn, when the company performance is given in terms of SAIDI, a simple, yet acceptable, way to calculate financial incentives is to multiply

- the SAIDI (in [minutes]), by
- the annual average load (total annual energy consumption/8760), expressed in [kW] and, of course, by
- the value of energy-not-supplied.[23]

In other words, the use of SAIDI implies a rougher approximation in the measure of the energy-not-supplied than is implied by the use of the ENS, which is calculated using load curves.

Nonetheless, measuring SAIDI is easier than calculating the ENS precisely because the latter requires information on load curves. Thus, choosing to regulate SAIDI instead of ENS can be a good solution, especially for newcomers in the field. An alternative solution is to derive a valuation of customer costs for interruptions in [euros/minute]. Financial incentives will thus be calculated multiplying the SAIDI by the number of customers served and by the per-minute cost of interruptions.[24]

Financial incentives aimed at the regulated company should reflect the costs sustained by the customers affected by the interruption. Hence, the accuracy in calculating financial incentives is high when one knows the cost of the energy-not-supplied (in [euros/kWh-not-supplied]) for a large number of customer groups, as well as the energy-not-supplied to each one of the same number of customer groups. This statement requires two specifications.

First, applied incentive schemes, including those regulating the ENS indicator (schemes that use load curves), generally limit the customer groups to a small number, especially at the initial stage. Regulators have used two, or even one,

[23] See the description of the Italian scheme in Section 4.4.5.

[24] See the description of the Dutch scheme in Section 4.4.5.

estimated load curves and correspondingly an equal number of different monetary valuations of the energy-not-supplied. According to the oft-mentioned principle of approaching quality regulation gradually, this number has been increased over time.[25]

Second, the SAIDI indicator for a distribution area does not carry any information about the type of user disconnected.[26] In practice, assuming the value of the energy-not-supplied for a given number of customer groups is known, the energy-not-supplied to each customer group still needs to be estimated *ex post*. For example, using the same approximation as above, the annual average load can be calculated for the different customer groups. In any case, given the same level of discretization (i.e., the same number of consumer groups), financial incentives will be a less accurate representation of consumer costs then when calculated on the basis of the ENS indicator.

In summary, both choices (the ENS and the SAIDI) lead to significant approximations with respect to the complexities of the real world. Accuracy in both cases is influenced, above all, by the level of discretization selected. A regulator should focus on introducing a simple and workable system. Part of the initial simplifications can be removed in time.

We observe now the form taken by the functional relationship between revenues and quality in applied incentive schemes.

The only European country implementing a simple linear incentive scheme, as depicted in Figure 4.8, is Norway. Indeed, Figure 4.8 is a synthetic representation of a more articulated scheme, as described below.

### The Norwegian scheme

A regression model is used in Norway to calculate the expected level of ENS (the performance standard) for each company $\mathcal{E}(ENS)$, in kWh. The model uses historical continuity data and includes various structural variables, such as energy supplied, network extension and weather conditions. Using the expected level of ENS and an average cost of interruption, $c_{m,n}$, differentiated for customer category, $m$, as well as for notified and un-notified interruptions, $n$, an expected Interruption Costs $\mathcal{E}(IC)$ is calculated, in euros, for each company:

$$\mathcal{E}(IC) = \sum_{m,n} \mathcal{E}(ENS)_{m,n} \cdot c_{m,n}$$

At the end of each year the regulator calculates the difference between expected and actual interruption costs, $IC$. If the difference is positive, i.e., if the quality of supply has been better than expected, the difference will be added to the company's revenue cap,

[25] For instance, in Portugal only one customer category is used (and one incentive rate); in Norway the two customer categories initially used were increased to six after two years, resulting in a total of twelve different incentives rates (the incentive rates are different for notified and un-notified interruptions).

[26] Greater accuracy could theoretically be achieved by measuring separate SAIDI for different voltage levels and/or geographical areas.

$R$. If the quality has been worse than expected the difference will be subtracted from the revenue cap. The variation in $R$ is thus:

$$\Delta R = \mathcal{E}(IC) - IC \tag{4.1}$$

Since distribution tariffs are set at the beginning of the year and the calculation in equation 4.1 is made *ex post*, there will be a deviation between allowed and actual revenues. This deviation is treated as an interest-bearing account either receivable from, or owed to, customers. This is either collected from customers through higher tariffs in subsequent years; or, if an amount is owed, it is paid back to customers through lower tariffs in subsequent years [Langset et al., 2001].

Incentive schemes in several countries (including Ireland, UK and The Netherlands) have a lower and an upper bound for penalties and rewards (Figure 4.9). Ideally, when financial incentives correctly reflect customer costs and benefits for quality variations, the introduction of upper and lower boundaries should not be necessary: the regulated company will reach the desired level of quality and stay there. However, as might be expected, the assumption of perfect information on customer marginal valuation of quality is not verified in practice. Thus, the risk of giving incentives for inappropriately high quality levels justifies the introduction of an upper boundary on rewards. Similarly, there is a risk of penalizing the company to the point where it will have unbearable financial losses. This supports the introduction, in applied mechanisms, of a lower boundary on penalties. In the UK and Ireland capping limits the percentage change in allowed revenues that is related to quality performance (respectively at ± 3% and ± 4% of price control revenues). In The Netherlands the change in revenues is capped at ±5%. The Dutch system is presented below.

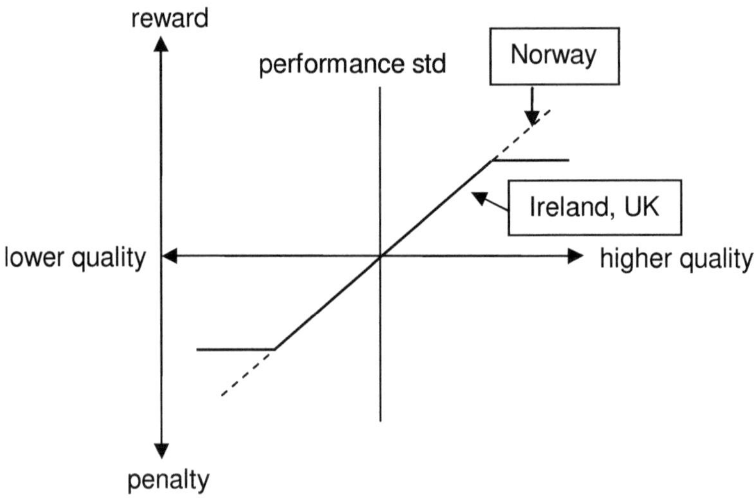


**Fig. 4.9.** Capping the incentive scheme

**The Dutch scheme**

The total revenues, $R$, of an electricity distributor, $i$, in a regulatory period depend on the change in the company *quality performance*, $\pi_i$, between two subsequent regulatory periods P1 and P2:

$$R_i = R_i^{-q} + (\pi_{i,P1} - \pi_{i,P2})$$

where $R_i^{-q}$ are the total revenues of company $i$ that include the adjustment for the retail price index and the efficiency gain (X factor) but not the adjustment for quality.

For a regulatory period of 3 years, starting in year $T$, the quality performance is calculated as follows. The difference between: the *company average SAIDI* and the performance standard is computed. The company average SAIDI, $\overline{SAIDI_i}$ is the average of the measured SAIDI over a three-year period (from the year $T-4$, to the year $T-2$). The performance standard, $SAIDI_{std}$ is the average SAIDI over all regulated companies and over a three-year period (from the year $T-7$, to the year $T-5$). The $\overline{SAIDI_i} - SAIDI_{std}$ difference is multiplied by the number of customers $N_i$ served by the company $i$ in the years $T-4$ to $T-2$ and the per-minute cost of interruptions over the same three years, $\varphi$. The result is increased by a *compensation allowance*, $CA_i$. Hence,

$$\pi_{i,P} = \sum_{k=T-4}^{T-2} N_{i,k} \cdot \varphi \cdot (SAIDI_{std} - \overline{SAIDI_i}) + CA_i$$

The per-minute cost of interruptions is calculated on the basis of two valuation functions, one for domestic customers, $d$, and one for non-domestic customers, $nd$. An analysis conducted by the regulator found a logarithmic relationship between the cost of quality and the number and duration of interruptions: $C^d(SAIFI, CAIDI)$, $C^{nd}(SAIFI, CAIDI)$.[27] The per-minute cost of interruptions, in $[\frac{euros}{min}]$, is thus determined in three steps: *(i)* replacing, in the two logarithmic functions, the average SAIFI and CAIDI over the years $T-7$, to $T-5$ and over all regulated companies; *(ii)* weighting the logarithmic functions by the percentage of domestic customers, *dom*, and non-domestic customers, *nondom*, to obtain a unique parameter; *(iii)* and dividing the latter by the average SAIDI over the years $T-7$ to $T-5$ and over all regulated companies.

The compensation allowance is determined by the annual number of customers compensated, $NC_{i,t}$ for an interruption that lasted longer than indicated in the Grid Code (4 hours) multiplied by the cost of this interruption (21 euros, in accordance with the above valuation functions, where $SAIFI = 1$, $CAIDI = 4$) and weighting over the percentage of domestic and non-domestic customers:

$$CA_i = \sum_{k=T-4}^{T-2} NC_{i,k}(C^d \cdot dom + C^{nd} \cdot nondom)$$

Note that the actual compensation that a distributor operator pays to low voltage customers amounts to 35 euros. Hence, the expenses related to the payment of compensation are only partially recovered by the regulated companies [DTe, 2002].

[27] CAIDI (Customer Average Interruption Duration Index) represents the average time required to restore service and is given as $CAIDI = \frac{SAIDI}{SAIFI}$.

It is important to note that another means to reduce the financial impact of the reward and penalty scheme is the exclusion of exceptional events from the calculation of the measured value of the regulated indicator. In practice, both capping and exemptions usually apply.

In addition to upper and lower boundaries, incentive structures can also include dead-bands (Figure 4.10). Dead-bands introduce an interval where rewards and penalties do not apply and are designed to avoid tariff variations for small deviations from the baseline. In fact, such deviations might not represent a structural change in quality, but simply a stochastic effect. A structure of this type is implemented in Portugal.

### The Portuguese scheme

The Portuguese incentive scheme affects the annual adjustments of allowed revenues for MV distribution networks activities. The performance standard (referred to in Portugal as the 'reference standard'), $ENS_{ref}$, is given in terms of a percentage of the energy supplied, $ES$, in a year and it is calculated every year: $ENS_{ref} = 0.0004 \cdot ES$ in 2005 and $ENS_{ref} = 0.00019 \cdot ES$ in 2006. A dead-band, $DB$, applies: $DB = 0.12 \cdot ENS_{ref}$. Under the incentive scheme three cases are given:

- When the annual measured energy-not-supplied $ENS_{mes}$ is lower than $ENS_{ref} - DB$ (indicating a good performance), the distributor's revenues are increased by an amount, $RQS$ (Revenues for Quality of Supply, in euros), equal to the difference between the two figures multiplied by an incentive rate, $V_{ENS}$ that expresses the value of the energy-not-supplied for a generic consumer:

$$RQS = V_{ENS} \cdot [(ENS_{ref} - DB) - ENS_{mes}]$$

- When the annual measured energy-not-supplied $ENS_{mes}$ is higher than $ENS_{ref} + DB$ (indicating a poor performance), the distributor's revenues are decreased by an amount, $RQS$, equal to the difference between the two figures multiplied by the incentive rate, $V_{ENS}$:

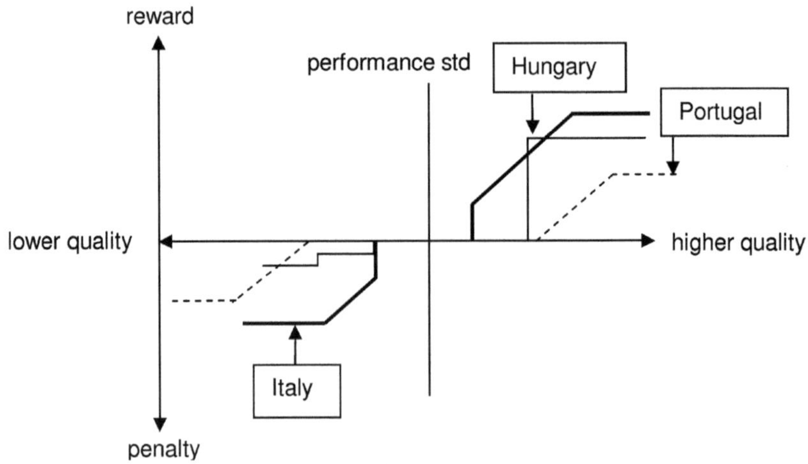


**Fig. 4.10.** Caps and dead-bands

$$RQS = V_{ENS} \cdot [ENS_{mes} - (ENS_{ref} + DB)]$$

- When the annual measured energy-not-supplied $ENS_{mes}$ is within the dead-band, the distributor's revenues are not affected.

In the year 2005 the value of the energy-not-supplied was 1.5 euros/kWh and the scheme upper and lower boundaries were set at 5 million euros. The change in revenues is applied with a two-year delay, i.e., in 2005 for the 2003 performance [CEER, 2005].

The Italian incentive scheme uses capping[28] and a dead-band (Figure 4.10), but it differs from the Portuguese system on two counts. First, the Italian dead-band (± 5% from the target) is much smaller than the Portuguese one (± 12%). Secondly, whereas schemes that are linear outside the boundaries of the dead-band (as in the Portuguese case) lead to small changes in tariffs when quality is just outside these boundaries, in the Italian scheme, once the change in quality is larger than the dead-band, the variation in the tariff is forced to be significant (note the vertical line in Figure 4.10). The idea is to avoid the administrative work implied in tariff modifications for small tariff variations (not only for small variations in quality). Note that the Norwegian scheme has no dead-band, yet the regulator requires companies to refrain from introducing changes in the tariff unless long-lasting changes in continuity have been achieved. The Italian incentive scheme is detailed below.

### The Italian scheme

The Italian reward and penalty scheme is applied to each of the almost 300 districts (administrative areas, defined as urban, suburban or rural, according to population density) that compose the national territory. Let us focus on a single district and assume that $S_i$ is the performance standard fixed by the regulator for the year $i$ (SAIDI in minutes). We define $A_i$ as the measured SAIDI indicator for the year $i$: this value is actually a two-year average of the measured SAIDI over the period ($i-1$, $i$), net of exemptions. $E_{i,1}$ is the total consumption (in kWh) of medium voltage users and low voltage non-domestic users in the district in year $i$; $E_{i,2}$ is the total consumption (in kWh) of low voltage domestic users in the district in year $i$. $UB$ and $LB$ are respectively the rewards and penalties upper and lower boundaries for the district. A dead-band, $db$, is defined as 5% of $S_i$. Incentives and penalties for the company serving the district are calculated according to the following formulae:

The case $A_i < S_i \cdot (1 - db)$ implies that the company has improved continuity more than required by the standard $S_i$; in this case the company gains a reward:

$$R_i = [max(S_i - A_i; UB)] \cdot \left(\frac{E_{i,1}}{8760} \cdot C_1 + \frac{E_{i,2}}{8760} \cdot C_2\right)$$

The case $A_i > S_i \cdot (1 + db)$ implies that the company has not improved continuity as required by the standard $S_i$; in this case the the company pays a penalty:

---

[28] As illustrated in Figure 4.9, the cap is twice as high for rewards as it is for penalties. It also varies by for territorial district.

$$P_i = [max(S_i - A_i; LB)] \cdot \left(\frac{E_{i,1}}{8760} \cdot C_1 + \frac{E_{i,2}}{8760} \cdot C_2\right)$$

The parameters $C_1$and $C_2$ are given in [euros/(minute kW)] and vary according to the district density and the measured SAIDI indicator, $A_i$. These values represent the average interruption costs for non-domestic and domestic users respectively, and they are higher (lower) for better (poorer) quality levels. The formulae used to define the performance standard, $S_i$, and the lower and upper boundaries, $UB$ and $LB$, are omitted here for the sake of brevity: they can be found in the official decision.

In Italy, the distribution tariff must be the same for all distribution companies. Hence, an equalization fund is used: penalties paid by utilities for under-performing districts are deposited in the fund and rewards due to utilities for over-performing districts are taken from the fund. The net difference between rewards and penalties modifies the Q factor in the quality-adjusted price cap formula:

$$p_t = p_{t-1} \cdot (1 + RPI - X \pm Q)$$

The $Q$ factor is calculated *ex-post* and can take a negative or a positive sign. When $Q$ is positive, it means that, as a whole, the system improved more than required and users are called on to contribute. On the contrary, when $Q$ is negative, it means that, as a whole, the system improved less than required and all users pay a reduced tariff [AEEG, 2004].

Finally, capping and dead-bands also apply to the Hungarian incentive scheme. The dead-band here is asymmetric: respectively, +10% for rewards and -5% for penalties. In addition, as explained below, the incentive scheme has the form of an asymmetric stepwise function (Figure 4.10).

### The Hungarian scheme

The Hungarian regulator has designed a system that controls three quality indicators: SAIDI, SAIFI, and network losses. As far as penalties are concerned, the distribution tariff is reduced

- by 0.5% when any of the three measured indicators is more than 5% worse than the performance standard but less than 10%;
- by 1% when any of the three measured indicators is more than 10% worse than the performance standard;
- When more than one measured indicator is worse than the performance standard, the penalties are added: in the worst case, a total 3% tariff decrease is applied.

Incentives are gained when at least one measured indicator is more than 10% better than the performance standard and the distributor is not subject to penalties regarding the other two regulated parameters. In this case the regulator allows the profit limit to be 10% higher. It should be noted that a three-year rolling average of the measured indicators is used in order to smooth out the volatility in quality performance that might be observed because of weather conditions [Tersztyanszky 2003; Tersztyanszky 2005].

It may be useful to summarize briefly the main ideas that emerged from the experiences described in this section.

- The description of the applied functional relationships between company revenues and continuity indicators highlights a diversity of approaches. This diversity derives from an adjustment of the regulation to specific industrial and institutional factors of the country and appears to contribute to its effectiveness.
- One basic common feature of these relationships is the presence of both penalties and rewards. The idea of an incentive scheme is to allow the regulated company to make relatively autonomous decisions regarding the level of quality delivered to its customers: for such decisions to be efficient, the company must be given clear signals about both customer costs for poor quality (penalties) and customer willingness to pay for better quality (rewards).
- In most incentive schemes (with the exception of Hungary) the relationship between quality and revenues is fundamentally linear: the same distance (in absolute value) from the performance standard leads to the same variation in revenues for the company. Note, however, that performance standards are not fixed over time. They usually change from one year to the next, requiring companies to meet progressively stricter performance standards.
- In the Hungarian system penalties take a different form: for any failure to comply with the performance standard there is a fixed reduction in the tariff. This approach has the advantage of avoiding the difficulty regarding the choice of the incentive rate. Still, the criterion to penalize the company more for larger deviations from the performance standard is maintained (the initial 0.5% tariff reduction per indicator increases, for worse performances, to 1%). The system is asymmetric and an incentive (a fixed increase in the profit limit) is gained only if performance is above a given threshold.
- Capping of the incentive scheme is necessary when one considers that the regulator has imperfect information about customer costs and willingness to pay. As such, it is highly recommended as part of the regulatory design.
- Dead-bands are most useful when they actually reduce the administrative work, and should be used for this purpose.
- Measuring performance against a two or three-year rolling average indicator helps capture stable changes in performance deriving from company investment and maintenance programmes. It also avoids the payment of penalties and rewards related to stochastic changes in continuity caused by exogenous parameters, such as weather conditions.
- Quality performance is normally regulated on a company to company basis. Hence, the distribution tariff paid by customers served by a company that performs well will be higher than the tariff paid in a poor-performing distribution territory. In some countries (Italy, for instance) there is a constraint requiring the distribution tariff to be the same across the national territory: the costs of quality are thus redistributed over the national customer basis.

We conclude this section with a practical observation. All regulatory instruments should be subject to *ex post* assessment. Incentive schemes should be assessed with respect to at least two variables: their effects on the regulated (and unregulated) quality indicators and their costs or benefits for companies and customers. The price control review is a good time to review the results achieved and to adjust the baseline in order to transfer to customers part of the benefits that derive from the quality improvements attained by the companies (see Figure 4.7).

## 4.5 Premium quality contracts

A fourth option for continuity of supply regulation is to allow distribution companies to offer individual quality contracts to customers. The experience with this instrument is currently quite limited, but its potential advantages are significant, for several reasons [Fumagalli et al., 2004].

First, customers have potentially very different needs in terms of quality of supply and different valuations of supply interruptions. Quality contracts allow, within certain limits, the delivery of a customized quality level and the payment of customized compensation in cases of failure to comply with the performance standards indicated in the contract. In general, customization is intended as a way to offer better quality levels than those offered with standard service (and not worse).

Second, premium contracts can be employed over different quality indicators: normally they would include continuity indicators as well as voltage quality parameters. Individual measures of the quality indicators included in the contract are obviously necessary: to date, this limits the use of contracts to medium-to-large customers. In addition, larger customers should possess the necessary information to assess the potential benefits of a premium quality contract against the costs incurred because of interruptions (and other voltage quality disturbances).

Third, the choices indicated in premium quality contracts reveal customer preferences and willingness to pay (WTP) for quality improvements directly to the provider of the distribution service. From the distributor's perspective, these are strong signals that can be taken into account in investment and maintenance decisions.

Finally, a regulator's control over premium quality contracts mainly concerns potential discrimination between customers with standard service and customized contracts. Otherwise, when contracts are used the role of the regulator is significantly smaller than when other regulatory instruments are implemented.

In turn, the main disadvantage faced by regulators and distribution companies (especially in case of an extension of quality contracts to LV customers) is the so-called 'free-riding' effect. Indeed, if a customer pays for better performance standards, the distribution company may well improve the network. This will also ensure a higher level of quality for all other customers connected to the same feeder. As a consequence, one customer pays for improvements that create benefits for many others. Further research is needed in order to determine what rules should be applied in cases of this sort.

Below we describe the French and Italian systems. A preliminary distinction is that in France continuity of supply is currently regulated using this instrument alone (a form of self-regulation adopted by the distribution company and supervised by the regulator),[29] whereas in Italy, premium quality contracts are an option that customers can choose on a voluntary basis: continuity of supply is regulated using all the other instruments described in this chapter.[30]

In France, a distribution company offers two types of contracts to eligible customers connected at MV level: a 'standard' contract and a 'customized' contract.

In the case of a standard contract, a distribution company takes on the obligation to provide less than a specified number of unplanned interruptions in a year. The number depends on the geographical area (urban, sub-urban or rural, in terms of population density), on the type of connection (with or without an alternative supply), and on the duration of the interruptions (long or short). The standard contract also fixes the number of planned interruptions: two interruptions per year, each lasting less than four hours. This contract makes the distributor responsible for damage directly caused to the customer in cases where the performance standards are not respected. Only direct and documented damages are taken into consideration and some exemptions apply.

Where a customized contract is concluded, a distribution company takes on the obligation to provide less than a computed number of unplanned interruptions in a year. The computation is based on a formula that takes into account the number of interruptions at the customer site during the previous four years. The contract requires the customer to pay an additional annual cost and obliges the distributor to install a monitoring device at the customer site. In case of a breach of contract the distributor is responsible for any damage caused to the customer. As before, only direct and documented damages are taken into consideration and some exemptions apply [ERD, 2000].

In Italy, the regulator establishes a few general rules to be observed by the distributors when offering premium quality contracts. The details of the contract are left to the interested parties to negotiate.

Premium quality contracts can include both continuity indicators, such as the duration and number of long and short interruptions, and voltage quality indicators.

A premium quality contract must indicate performance standards for one or more quality indicators, a premium price to be paid by the customer, compensation to be paid by the distributors in cases of non-compliance with the performance standard(s), and the events that are excluded from such payment.

Individual performance standards cannot be lower than those defined by the regulator for the same indicators at system-level.

For quality indicators that are not already measured on an individual basis, the parties are obliged to arrange for a period of measurement (not less than one year) prior to the stipulation of the contract. The cost of such measures is born by

[29] Other regulatory instruments are currently being examined by the regulatory authority.

[30] Note, however, that to date no premium quality contracts have been concluded in Italy.

the customer who is entitled to install its own measuring instrument [AEEG, 2004, Lo Schiavo et al., 2005].

## 4.6 Advanced topics

In continuity of supply regulation there are areas where new solutions are proposed for traditional problems and areas that are now receiving renewed attention. Without entering into the details of all these areas, it is important to summarize them here.

A first area of interest regards the treatment of exceptional events.

This issue has proven to be one of the most troubling for regulators in practice. Definitions of exceptional events or *force majeure* have proved not to be sufficiently equitable or unambiguous when applied to the different situations that have occurred in practice. These events occur with a low frequency but also have potentially substantial social and economic consequences, as well as a significant impact on the amount of penalties paid by the distribution company. Given their importance, alternative solutions have been proposed for identifying exceptional events and for handling the relative continuity data separately from normal operation data. These innovative approaches are based on statistical methodologies: the underlying idea is that exceptional events are located on the right-hand tail of the distribution of daily SAIDI.

A statistical method has several advantages because it: simplifies administrative procedures; reduces the costs incurred by companies and by the regulator in order to provide and control the correct exclusion of the event; statistically distributes the errors of the identification method; and is easy to understand (only simple statistical concepts such as averages and standard deviations are used). However, the statistical analysis of continuity indicators is a relatively unexplored area, and several methodological issues remain open. Different methodologies have been conceived: one that has been proposed in the US, described in IEEE (2004); one that has been applied in the UK [Ofgem, 2004a]; and one that has been applied in Italy [Fumagalli et al., 2006, Salvati et al., 2006].

The exclusion of exceptional events from the incentive regulation does not mean that the problem is receiving less attention. On the contrary, several steps have been taken towards introducing procedures for managing emergencies. The UK regulatory authority, Ofgem, has introduced a regulatory framework to provide incentives for a prompt operational response even in cases of severe weather conditions. This interesting methodology is described in [Ofgem, 2004a].

A second area of renewed interest regards network reliability in the medium-term. The reliability of the distribution network is defined by a gradual process that involves, on the one hand, the deterioration of the network due to ageing and wear and, on the other hand, the improvement of the network by replacement of poorly performing assets. Whilst it is quite clear that regulation positively affects the regulated quality indicators, little information is available on precisely how these results are achieved, and there may be a risk of encouraging short-term improvements, at the potential expense of medium-term network performance. Indeed, regulated indicators of quality performance may not be a satisfactory measure of the above process. Concerns on medium-term network performance have generated an interesting debate on the role and future direction of continuity regulation.

A first type of response to the problem would be to monitor quality expenditure. Reviewing the proposed level of expenditure or even the detailed expenditure by equipment type or function requires the regulator to have significant knowledge of the relationship between expenditure and network quality. In any case, finding an efficient point in expenditures could also prove difficult for the regulator. A second, more interesting group of proposals points in the direction of a regulatory approach which, instead of looking at quality indicators, becomes increasingly involved with the process behind them. In other words, it envisages a regulator who examines the details of fault statistics and asset management. Ofgem has recently introduced regulatory instructions and guidance to monitor the medium-term performance of network assets (i.e., analyzing fault as well as ageing and replacement statistics of network components)[Ofgem, 2005b].

Finally, there is growing interest in extending the same regulatory instruments employed today in distribution to the performance of the transmission network. Currently, however, there are only a few countries (including France, Portugal and Hungary) where individual standards of performance (and compensation payments) have been introduced for transmission users (i.e. distributors, customers directly connected to the transmission grid and production plants).

In practice, the continuous evolution of continuity regulation provides interesting indications for regulators approaching these issues for the first time. As such, international progress in quality regulation should be closely monitored. Regulators should be careful, however, to approach problems incrementally and to prefer solutions that are more compatible with the specific national environment: the availability of reliable measurements, the industry structure, the initial level of quality, and the institutional framework. Regulators in different countries achieve desired outcomes using different methodologies and shaping regulation to local conditions is a key factor in the success of existing regulatory frameworks.

# 5

# Voltage quality

Voltage quality regulation focuses on a subset of a wide range of factors that describe the deviations of the voltage from its ideal waveform. Such deviations, or disturbances, can provoke damage to or malfunctioning of customers' electrical equipments (or cause annoyance to people - e.g., visual annoyance).

Conceptually, the regulation of voltage quality could make use of all four regulatory instruments described in Chapter 2. However, voltage quality regulation is still quite rare in Europe. The availability of reliable measures is indeed a very critical issue: sometimes there are not even universally accepted definitions of specific disturbances. Several countries have introduced voltage quality standards and premium quality contracts, but in most cases these instruments do not yet constitute a comprehensive regulatory approach to voltage quality problems. In turn, the interest of regulators in voltage quality has recently increased: there are many reasons for this trend, including the growing sensitivity of end-user equipment due to the wide use of electronic devices in every electrical application.

Section 5.1 provides brief definitions for the main voltage disturbances and discusses the subject of data collection from a regulatory perspective. Section 5.2 deals with publication of data, Section 5.3 with minimum quality standards and Section 5.4 with premium quality contracts. Reward and penalty schemes have not been employed in voltage quality regulation. For this reason the subject is not treated here. A few advanced topics are discussed in Section 5.5.

It is important to state clearly that voltage quality is a very technical and specific subject. In addition, many of the issues related to voltage quality are already addressed by technical norms which define the voltage characteristics of the electricity supplied by distribution companies and which are normally employed as a reference. Technical norms are issued by international electrotechnical committees (not by regulators) for standardization purposes.[1] Defining characteristics of electricity

[1] Technical norms are often referred to as *standards*. In this handbook we intentionally use *technical norms* in order to avoid confusion with *performance standards*, which are used in regulatory instruments and are defined by regulators.

supply is necessary in order to facilitate circulation of electrical appliances and avoiding interference between electrical systems and customer equipment.

The fundamental reference for voltage quality in Europe is the European Norm EN 50160, issued by the European Committee for Electrotechnical Standardization (CENELEC) in 1994.[2] It was then revised in 1999, as required by EU Directive 85/374/EEC on liability for defective products (which includes electricity) [EC, 1985]. Another important reference is the IEC 61000 series of technical norms on electromagnetic compatibility (EMC) issued by the International Electrotechnical Commission (IEC).[3] These norms concern voltage disturbances, immunity and emissions from the electrical equipment, and they describe measurements and measurement techniques for voltage quality characteristics.

Due to its complexity, voltage quality is not an issue for regulators approaching service quality regulation for the first time. Nonetheless, voltage quality regulation is included in this handbook for two reasons.

First of all, an increasing number of customers suffer from voltage disturbances. Regulators are likely to be involved in investigations, in providing responses to customer complaints and in other regulatory activities that require a basic understanding of voltage quality issues. Thus, this handbook is intended to be not so much a technical reference on the subject (for which other textbooks are available), but as a guide to understanding the challenges that voltage quality introduces for regulators.

In addition, several European regulators find that the EN 50160 is not completely satisfactory for customer protection. To date, only a small number of countries in Europe implement voltage quality regulation (some regulators do not have legal power on this issue). However, for several regulators monitoring voltage quality and looking for better solutions to customers' problems than those offered by the EN 50160 is becoming an important activity. Newcomers in the field are thus encouraged to be aware of the importance of the subject and to include voltage quality in their medium to long term plans for service quality regulation.

## 5.1 Definitions and data collection

Describing voltage disturbances and the issues related to their measurement is an extremely technical task. In addition, a consolidated regulatory experience on voltage quality data collection is not available. Thus, the objective of this section is to provide only the few basic notions on voltage disturbances that are necessary to understand the challenges posed by voltage quality regulation. This handbook

[2] CENELEC (www.cenelec.org) works in the interest of European harmonization, creating technical norms either requested by the market or in support of European legislation, helping to shape the European Internal Market.

[3] IEC (www.iec.ch) is the leading global organization which prepares and publishes international technical norms for all electrical, electronic and related technologies. CENELEC has adopted several IEC technical norms, which also function as CENELEC technical norms.

is not meant to be a substitute for a good power quality textbook on the subject. For the same reason, we will provide only a few general indications on measuring voltage quality, giving precedence to those that are most relevant from a regulatory perspective. In particular, after a short general introduction, we present some technical definitions in Section 5.1.1 and a few indications regarding data collection in Section 5.1.2.

The characteristics of supply voltage (frequency, magnitude, waveform, symmetry of the three phase voltages) are subject to variations during the normal operation of an electrical system [CENELEC, 1999]. These deviations from the nominal or desired values are called *voltage disturbances*, and can be grouped into two categories [Bollen, 2000]:

- *Voltage variations*: small deviations from the nominal or desired value. A voltage variation is characterized by having a (different) value at any moment in time. To measure the characteristics of a voltage variation (for instance, the frequency of voltage fluctuations), an uninterrupted monitoring is necessary. Voltage variations are mainly due to load pattern, changes of load, or nonlinear loads. Supply voltage variations, voltage fluctuations leading to flickers, voltage unbalances and voltage harmonics are all examples of voltage variations.
- *Voltage events*: sudden and significant deviations from the nominal or desired values. Voltage events only happen every once in a while. To measure the characteristics of a voltage event (for instance, the depth of a voltage dip), a triggering mechanism can be used. This mechanism starts recording when the relevant characteristics exceed a given threshold. Interruptions (a sudden drop of the voltage to zero) are the best-known voltage events. Rapid voltage changes, voltage dips, swells and transient over-voltages are other important voltage events.

The distinction between voltage variations and voltage events is particularly relevant for a regulator.

Voltage variations are the 'physiology' of network functioning: it is impossible for a voltage characteristic to be equal to its nominal value at all times. However, customer equipment is designed to work optimally with an ideal voltage waveform. In addition, voltage variations outside predefined limits may lead to severe problems for customers. Distribution network operators can do a lot to keep voltage variations as small as possible. This eventually leads to a more efficient network management. For instance, keeping the magnitude of the supply voltage close to its nominal value is related with having fewer electricity losses. Further improvements in this sense can be achieved if customers maintain their power factors close to 1, and distributors can have an active role in this.[4]

Voltage events represent the 'pathology' of network functioning: they are by nature occasional and, fortunately, rather rare. However, they are also of large concern

[4] This matter is part of the larger subject of power quality. However, it is beyond the scope of this handbook.

for end-user equipment. Voltage events can bring an industrial process to a complete stop (even when there is no interruption of the supply). Distribution network operators can prevent and mitigate voltage events, but they cannot eliminate them completely. Therefore, customers using equipment which is particularly sensitive to voltage events should consider taking action: there are technical solutions that enable a sensitive customer to override certain voltage events with negligible or no consequences for the equipment.

A second aspect of voltage quality disturbances that is particularly relevant for regulators is the fact that most of them are the result of the interaction between customers and the network. A typical example of this interaction is the harmonic distortion of the voltage waveform. The presence of non-linear loads (e.g., power electronics devices, rectifiers, and so forth) gives rise to harmonic currents; the value of such currents depends on the network impedance (i.e., on the short-circuit power at the point of connection); as a consequence, the disturbance perceived by the other customers located along the same feeder depends both on the disturbance introduced by non-linear loads and on the network characteristics.

### 5.1.1 Voltage disturbances: definitions

For the purposes of this handbook, we briefly describe here the following phenomena:

- supply voltage variation;
- rapid voltage change;
- voltage dipand swell;
- flicker;
- voltage unbalance;
- harmonic voltage distortion;
- transient over-voltage;
- mains signalling voltage.

The descriptions given here are not meant to be exhaustive. We provide a synthetic definition of the disturbances, the typical cause(s), the main consequences for customers and a few indications regarding measurement rules. In most cases we report the definitions of voltage disturbances given in the EN 50160 for low and medium voltage networks [CENELEC, 1999]. Note that low voltage (LV) is defined in the EN 50160 as a voltage whose upper limit is 1 kV, and medium voltage (MV) is defined as a voltage between 1 kV and 35 kV. We refer the interested reader to the engineering literature for a comprehensive and more precise introduction to the vast field of voltage quality.

#### Supply voltage variation

In an electrical system, the supply voltage (i.e., the voltage at the point of connection of the customer's installation to the distribution system) should ideally be equal to a nominal level, $U_n$. A voltage variation is an increase or decrease of the voltage magnitude with respect to the nominal level, typically due to variations of the load.

The steady state voltage magnitude is a fundamental for the operation of electrical equipment, which is designed to work properly at the nominal voltage. This characteristic of the voltage depends on system design characteristics, load changes and other changes to which the system is subject. It is common practice to correct steady state voltage at various points of the system, for instance, using automatic tap changers at transformer substations. In particular, rural customers can experience significant supply voltage variations.

The relevant indicator for this disturbance is the magnitude of the supply voltage, evaluated as the mean of the root-mean-square (rms) values[5] of the voltage in a given time interval. According to the EN 50160, the time interval is 10 minutes. One important criticism that has been aimed at the EN 50160 is that a 10-minute mean can hide important variations. Using a shorter time interval would be more precise. Figure 5.1 illustrates the voltage magnitude evaluated as a 10-minute and 1-minute mean, simultaneously for the same point of connection to the distribution network. For a nominal voltage of 230 V, the 10-minute mean is lower than 207 V (-10% of the $U_n$) 3.5% of the time. Instead, the 1-minute mean is below 207 V 28 % of the time.

In some European countries, shorter time intervals are used. Norway uses a 1-minute mean normally. Hungary uses a 1-minute mean only when the maximum level of voltage needs to be verified.

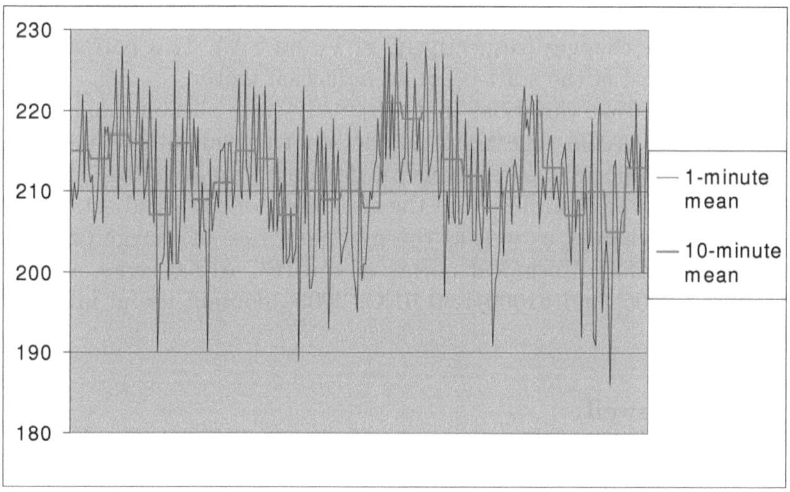


**Fig. 5.1.** Supply voltage variations [ERGEG, 2006]

[5] The rms (root-mean-square) value is the square root of the arithmetic mean of the squares of the instantaneous values of a quantity, taken over a specified period of time.

## Rapid voltage change

A rapid voltage change is defined in the EN 50160 as "a single rapid variation of the rms value of a voltage between two consecutive levels which are sustained for definite but unspecified durations". Load switching, transformer tap-changers, and switching actions in the electrical system can lead to a sudden change in voltage magnitude. Normally, both voltage before and after the step are in the normal operating range (typically 90% to 110% of the nominal voltage) [Bollen, 2000]. This means also that rapid voltage changes are classified as different phenomena from voltage dips and swells (see *infra*).

Rapid voltage changes within this range will mainly lead to visual annoyance, and should not cause any damage or malfunction of electrical equipment (although this is still a field for further research).

Rapid voltage changes present different characteristics when caused by different events. These characteristics include the 'maximum voltage change' (the difference between the initial steady state and the lowest or highest voltage level during the event) and 'the steady-state voltage change' (the difference between the initial steady-state voltage level and the final steady-state voltage level), both typically expressed in percentages of the nominal voltage.

Figure 5.2a shows an instantaneous change where the maximum and the steady-state voltage changes are the same. The maximum voltage change is 10 V and the steady-state voltage change is 10 V. Figure 5.2b shows a gradual change. Note that the voltage change in time ($dv/dt$) is different than before; however, the maximum and the steady-state voltage changes are the still same (10 V). By contrast, the rapid voltage change illustrated in Figure 5.2c presents different maximum and steady-state voltage changes (respectively, 27 V and 5 V). This pattern (exponential change) is typical of the start-up of an induction motor.

Unlike supply voltage variations that are steady-state phenomena, rapid voltage changes are voltage events. The EN 50160 does not provide precise indications for identifying them. Important parameters for defining a rapid voltage change would include indicators of the duration and the steadiness of the steady-state condition before the voltage change, as well as the minimum rate of change ($dv/dt$) of the voltage [ERGEG, 2006]. Technical norms of the IEC 61000 series, such as IEC 61000-3-3 [IEC, 2005] and 61000-4-30 [IEC, 2003], contain useful information in this regard.

## Voltage dip and swell

A voltage dip (or, especially in American parlance, 'voltage sag') is a reduction in the supply voltage magnitude followed by a voltage recovery after a short period of time. The duration of the voltage reduction can go from a cycle[6] to a second (mostly from tens of milliseconds to hundreds of milliseconds).

[6] In Europe the frequency of the supply voltage is 50 Hz. The duration of one cycle is thus 20 ms (1 cycle = 0.02 s).

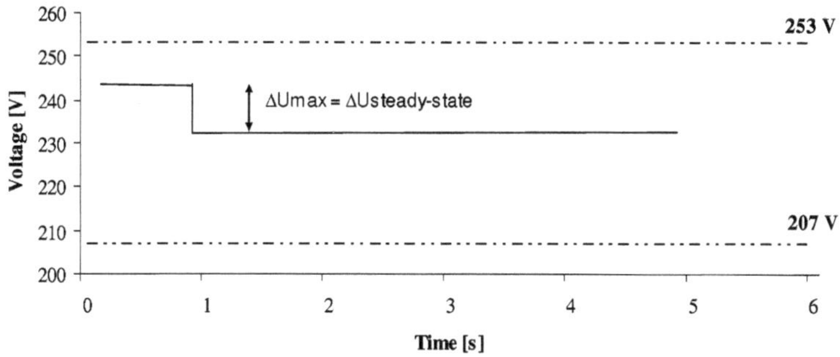


(a) Instantaneous change

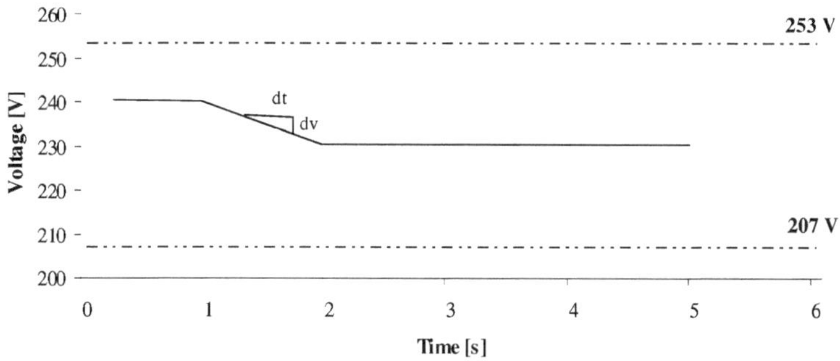


(b) Gradual change

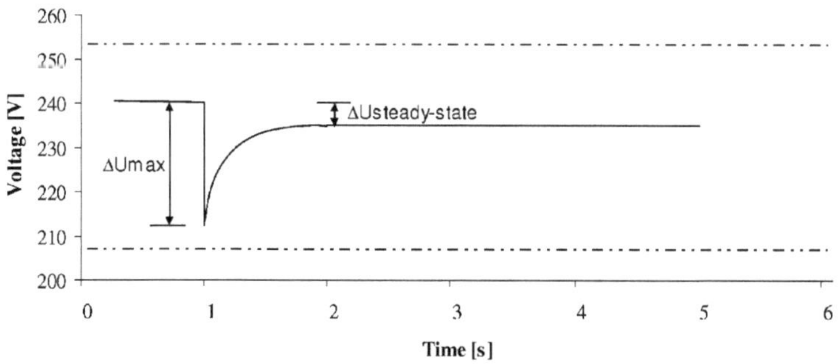


(c) Exponential change

**Fig. 5.2.** Rapid voltage changes [ERGEG, 2006]

A voltage swell is the opposite phenomenon: an increase in voltage magnitude followed by a voltage recovery, over the same time range. Voltage swells are also called 'temporary' over-voltages and are different from 'transient' over-voltages, which normally last less than half a cycle (10 ms).

As far as voltage dips are concerned, it is worth remembering that every branch of a power system is characterized by an impedance; thus, every increase in the value of the current flowing in that branch causes a corresponding reduction in voltage. Usually, these reductions are small enough that the voltage remains within normal tolerances. When there is a large increase in current, or when the system impedance is high, the voltage can drop significantly. Practically, the most common causes of voltage dips are faults or short-circuits and the starting of large load.

Voltage swells originate from, among other causes, lightning strokes, switching operations, and sudden load reduction.

According to the EN 50160, the duration of a voltage dip is conventionally set between 10 ms and 1 minute. The depth of a voltage dip is defined as the difference between the minimum rms voltage during the voltage dip and the declared voltage.[7] A sudden reduction of the voltage is recorded as a voltage dip only when the supply voltage is reduced to less than a given threshold, set conventionally at 90% of the declared voltage. Similarly, the threshold is conventionally set at $U_n$+10% for swells.

As illustrated in Figures 5.3 and 5.4, voltage dips and swells are characterized through their depth (or height) and duration.

The depth and the duration of a voltage dip and the height and duration of a voltage swell are of primary importance in terms of consequences for the customers. This is why dips and swell are typically described using tables and charts with two dimensions (not a single indicator).

For instance, it is common to collect data on the frequency of voltage dips using a table with magnitude and duration ranges. As illustrated in Table 5.1, each element in the table gives the number of events in a period of time, with magnitude and duration within a certain range, e.g., magnitude between 40% and 70% of the nominal voltage and duration between 200 and 500 ms.

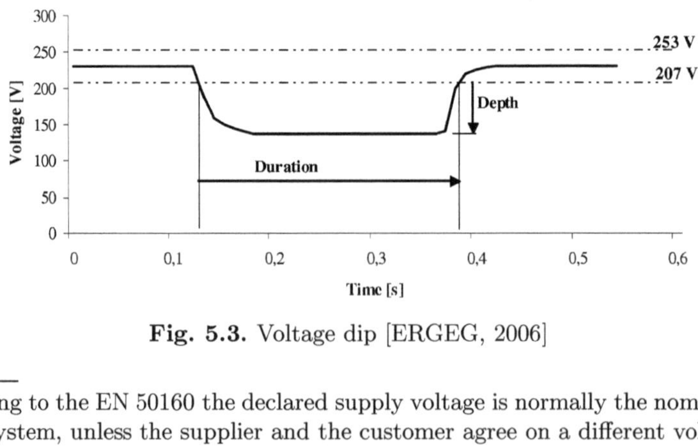


**Fig. 5.3.** Voltage dip [ERGEG, 2006]

[7] According to the EN 50160 the declared supply voltage is normally the nominal voltage of the system, unless the supplier and the customer agree on a different voltage.

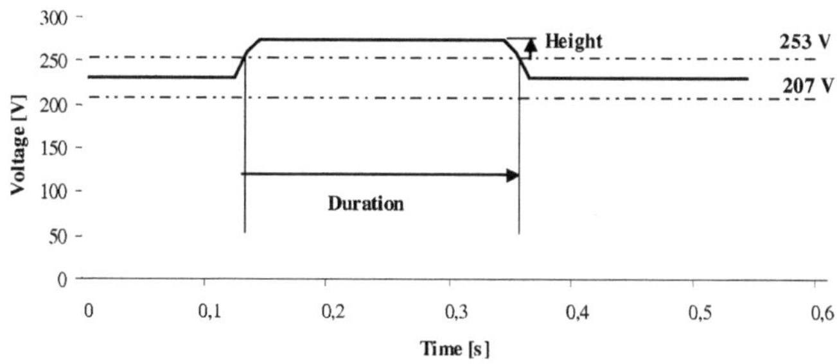


**Fig. 5.4.** Voltage swell [ERGEG, 2006]

The consequences of voltage dips and swells on end-user equipment are generally represented with 'voltage tolerance' curves, such as the ITI Curve (published by the Information Technology Industry Council) for the information technology products, which will be discussed in Section 5.5 [ITIC, 2000].

It is important to note that the number and depth of voltage dips measured at a point on the distribution network depend on a number of factors, including:

- extension and meshing of the HV network;
- extension of the MV network fed by the substation;
- presence of cable or overhead lines;
- neutral grounding (ungrounded, resonant, solidly grounded);
- short-circuit power;
- behaviour of end-users (motor starting, inrush currents);
- distributed generation.

### Flicker

A voltage fluctuation is defined by the EN 50160 as a series of voltage changes or a cyclic variation of the voltage envelope. Voltage fluctuations cause changes of the luminance of lamps, which can create the visual phenomena called 'flicker': the visual discomfort caused by repetitive changes of brightness in lighting.

**Table 5.1.** Frequency of voltage dips in (one year) and for a given point of connection (numbers are for illustration only)

| | Duration $t$ [ms] | | | | | |
|---|---|---|---|---|---|---|
| Residual voltage $U$ [%] | $10 < t \le 20$ | $20 < t \le 200$ | $200 < t \le 500$ | $500 < t \le 1000$ | $1000 < t \le 5000$ | $5000 < t \le 6000$ |
| $90 > U \ge 80$ | 7 | 11 | 2 | 1 | 1 | 0 |
| $80 > U \ge 70$ | 0 | 9 | 2 | 1 | 0 | 0 |
| $70 > U \ge 40$ | 0 | 48 | 5 | 1 | 0 | 0 |
| $40 > U \ge 10$ | 0 | 18 | 2 | 0 | 0 | 0 |
| $10 > U$ | 0 | 1 | 0 | 0 | 0 | 0 |

Voltage fluctuations can give rise to flicker for frequencies of the voltage envelope up to 25 Hz. Figure 5.5 shows an example of such voltage fluctuations where the frequency of the voltage envelope is 12.5 Hz. Phenomena of this type are typically caused by certain industrial loads such as welding machines, rolling mills, large motors with variable loads, and arc furnaces.

The visual annoyance caused by flickers grows very rapidly with the amplitude of the fluctuation. For the most part, however, equipment is not disturbed by this phenomenon.

To quantify the intensity of flicker annoyance, two indicators of 'flicker severity' have been introduced. The first one captures short term severity, $P_{st}$ (Perception of flicker, short term), and it is measured over a period of ten minutes.[8] The second one captures long term severity, $P_{lt}$ (Perception of flicker, long term)[9] and it is calculated over a two hour interval [IEC, 2003].

## Voltage unbalance

According to the EN 50160, in a three-phase electrical supply system, a voltage unbalance is a condition in which the rms values of the phase (or line) voltages, or the phase angles between consecutive phases voltages, are not all equal (see Figure 5.6). Voltage unbalances have two main causes: asymmetries of line impedance and load unbalance.

The main negative effect of voltage unbalance is on three-phase induction motors (one of the most common loads on the network and especially common in industrial

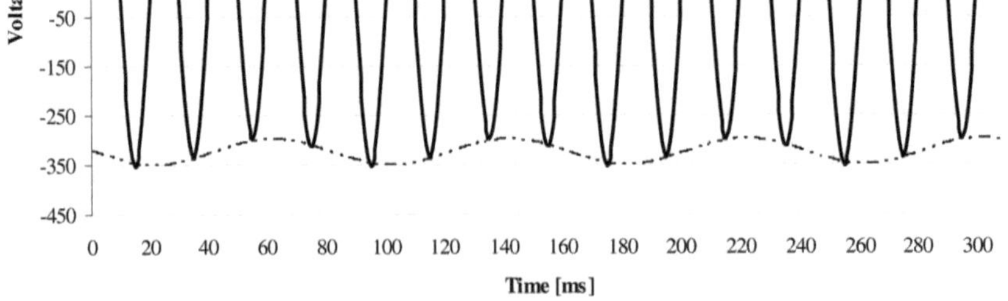


**Fig. 5.5.** Example of voltage fluctuations giving rise to flicker [ERGEG, 2006]

8 $P_{st}$ characterizes the likelihood that the voltage fluctuations will result in perceptible light flicker. A value of 1.0 represents the level that will result in 50% of people perceiving flicker in a 60 W incandescent bulb.

9 $P_{lt}$ is calculated according to the following expression: $P_{lt} = \sqrt[3]{\sum_{i=1}^{12} \frac{P_{sti}}{12}}$

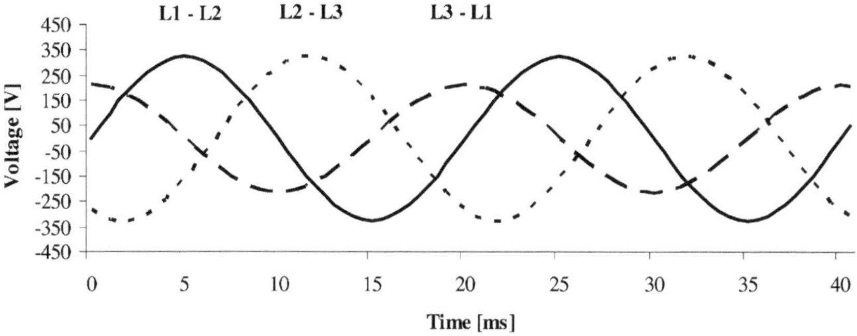


**Fig. 5.6.** Supply voltage unbalance in a three-phase system [ERGEG, 2006]

environments). In particular, voltage unbalance degrades the performance of such motors and shortens their life.

The level of voltage unbalance present in a system is specified in the technical norms in terms of values of the negative sequence component of the supply voltage.[10] This component is in fact the relevant one for the possible interference of appliances connected to the system. In particular, the voltage unbalance indicator is the ratio of the 10-minute mean rms values of the negative phase sequence component of the supply voltage to the positive phase sequence component [IEC, 2003].

### Harmonic voltage distortion

The voltage waveform is never exactly a single-frequency sine wave. This phenomenon is called 'harmonic voltage distortion'. A periodic waveform can be described as a sum of sine waves with frequencies that are multiples of the fundamental frequency. The non-fundamental components are called 'harmonic distortions' [Bollen, 2000].

The EN 50160 notes that harmonics of the supply voltage are caused mainly by customers' non-linear loads connected to all voltage levels of the supply system.[11] Harmonic currents flowing through the system impedance give rise to harmonic voltages. Harmonic currents and system impedance, and thus the harmonic voltages at the supply terminal, vary in time.

Many devices on the power system respond poorly to non-sinusoidal waveforms. Transformers, for example, become less efficient. Many energy meters become less accurate. Protection devices such as circuit breakers may trip too soon (unwanted

[10] According to the theory of Symmetrical Components, an unbalanced system can be broken down, mathematically, into three balanced systems, called positive sequence, negative and zero sequence systems. For a balanced system both negative and zero sequence systems would be absent.

[11] For instance, loads fed through power-electronic converters draw a non-sinusoidal current: the harmonic current components cause harmonic voltage components, and thus a non-sinusoidal voltage in the system [Bollen, 2000].

trips), or too late. Balanced harmonics at multiples of 3 of the fundamental freqvency, or triple harmonics (3rd, 9th, 15th, etc.), fail to rotate on three-phase systems: as a result, neutral conductors may overheat.

Harmonic voltages are defined in the EN 50160 as sinusoidal voltages with a frequency equal to an integer multiple of the fundamental frequency of the supply voltage (50 Hz). The measurement of harmonic voltages requires sophisticated equipment, especially for harmonics of higher order.

Harmonic voltages are evaluated, according to the EN 50160, by means of two indicators:

- Individually by their relative amplitude ($u_h$) related to the fundamental voltage $U_1$, where $h$ is the order of the harmonic;
- Globally, by an index of harmonic distortion, the total harmonic distortion factor THD, calculated as:

$$THD = \sqrt{\sum_{h=2}^{40} (u_h)^2}$$

Figure 5.7 illustrates a distorted sine wave and the individual harmonic components. The total harmonic distortion of the depicted voltage waveform, THD, is 7.7%.

Harmonic voltages are a permanent disturbance: all relevant parameters are measured over a 10-minute time interval. Harmonic voltage levels are evaluated excluding periods with fast transients, voltage dips, temporary over-voltages, and interruptions.

### Transient over-voltage

Transient over-voltages are voltage disturbances of very short duration (up to a few milliseconds) but high magnitude (up to several thousand volts, kV) with a very fast rise time (from milliseconds down to much less than a microsecond). Figure 5.8 illustrates an example of transient over-voltages. These disturbances arise because of lightning strokes and switching of heavy or reactive loads.

Because of the high frequencies involved they are considerably attenuated as they propagate through the network, so that those transient over-voltages occurring close to the point of interest will be much larger than those originating farther away. Protective devices in the network (surge arresters) ensure that such disturbances are generally kept to a safe level, and most problems arise because the source of the transient over-voltage is close to or within the installation.

The damage that results may be instantaneous, such as the catastrophic failure of the electrical plant or appliances, or the corruption of data within computers or on network cabling. Damage may also be progressive, with each event doing a little more damage to insulation materials until a catastrophic failure occurs.

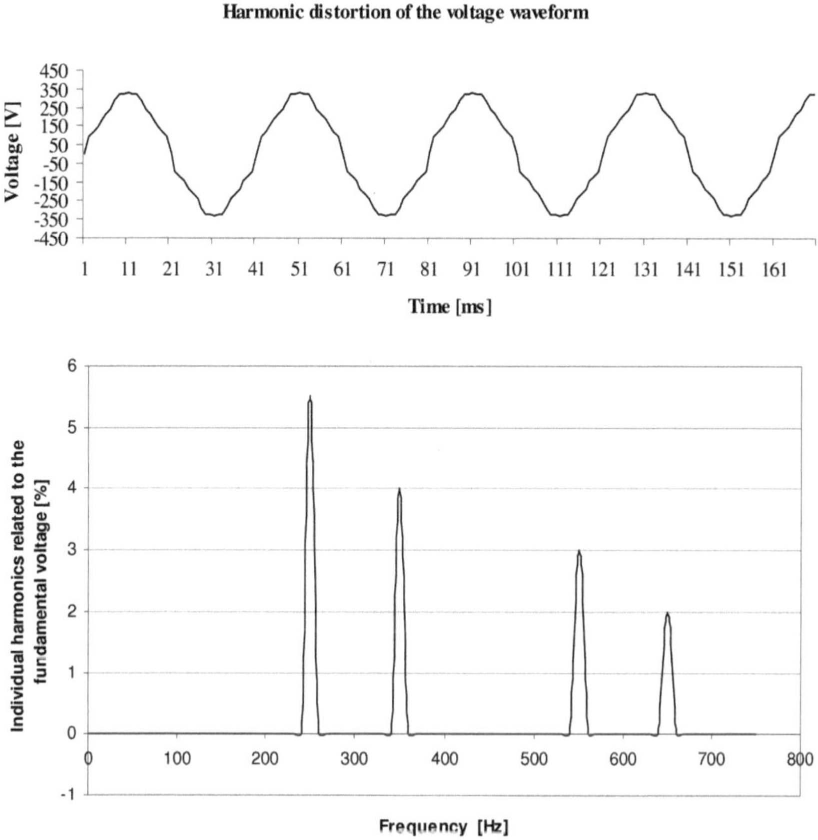


**Fig. 5.7.** Example of a distorted voltage [ERGEG, 2006]

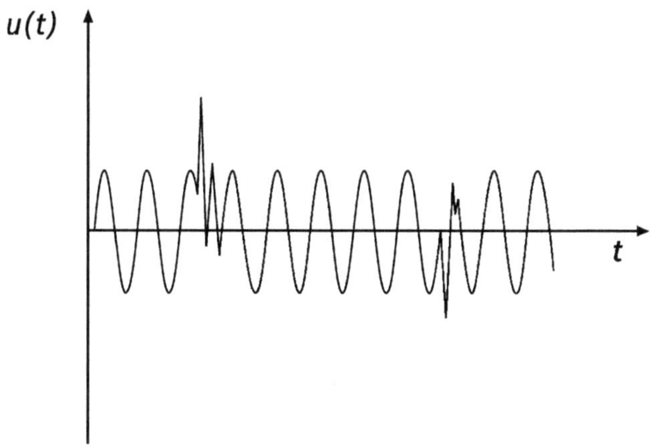


**Fig. 5.8.** Example of transient over-voltages [ERGEG, 2006]

### Mains signalling voltage

The distribution network can be used for the transmission of signals. According to the EN 50160, the mains signalling voltage is a signal superimposed on the supply voltage for the purpose of transmission of information through the distribution system and to customers' premises. Mains signalling voltage can interfere with equipment using similar frequencies for some internal purpose.

The relevant indicator is the three-second mean of signal voltages. These values must be kept within specified limits (see Section 5.3).

## 5.1.2 Data collection

From a regulatory perspective, the basic information on data collection for voltage quality include the following.

First, unlike interruptions that concern large numbers of customers at the same time, voltage characteristics vary randomly, over time in a specific network point, or by location at a given point in time. Thus, the level of voltage quality experienced by customers needs to be measured at their point of connection to the distribution network.

Second, measuring voltage quality disturbances requires the installation of a specific voltage quality recorder (voltage characteristics cannot be obtained using normal SCADA systems).

Third, the distinction between voltage variations and voltage events is relevant. Voltage variations are rather easily detectable by an occasional measurement, as those disturbances are always present on distribution networks. A two or three-week monitoring period is often enough to ascertain the importance of voltage variations at a given point on the network. In turn, voltage events occur rarely and at different point on the distribution system. Thus, in order to detect voltage events, a continuous monitoring is necessary.

Fourth, as of today, measuring voltage quality for each and every customer on a distribution network is not economically feasible. Nonetheless, it is extremely interesting for regulators to acquire a general understanding of the actual levels of voltage quality. This can be achieved by measuring voltage quality on a given number of points on the network. Along with system-level measurements, individual measurements (at the customer point of connection) are necessary when a problem is detected.

Fifth, as of today it is extremely rare for regulators to have introduced their own regulatory instructions and guidance on voltage quality measurement. This regards both the measuring methodology and the obligation to measure voltage quality. Typically, measurement methodologies coincide with those indicated in technical norms. A few exceptions apply, e.g., the requirement (mentioned in Section 5.1.1), introduced in Norway, to measure voltage variations using a 1-minute mean rms instead of a 10-minute mean.

As far as system-level measurements are concerned, mandatory monitoring of voltage quality on the distribution network is not common. However, there are a few exceptions, including the following.

- In Hungary, the regulator owns 400 quality recorders that are installed each semester in one of the 6 distribution companies on the LV network. The regulator chooses the network points randomly, in a way that does not depend on previous events or complaints [CEER, 2005].
- Since 2006, the Norwegian regulator has introduced a mandatory requirement for each network company to monitor voltage quality continuously on their networks in order to detect voltage events (dips, swells and rapid voltage changes). In particular, companies have to identify a number of representative measurement points, taking into account a number of network characteristics including grid extension, type of grounding of the system, percentage of cable and/or overhead lines, type of connected customers and climate conditions. Voltage variations should also be measured. In this case, one to two weeks of measures are considered satisfactory, although voltage variations may be different in summer/winter, week days/weekends [Brekke, 2006].

An alternative approach to mandatory monitoring is a measuring campaign. In recent years, several European countries have installed measuring systems, often with the commitment or the involvement of regulators, and other countries are planning to do the same (for additional information, see CEER, 2005, Section 4.2). The monitoring systems implemented present different characteristics in terms of number of quality recorders installed, the duration of the measurements, and the cost recovery method. In turn, these monitoring systems generally collect similar data. The interest is focused on long and short interruptions, supply voltage variations, voltage dips and voltage harmonics. The idea is that, in addition to contributing to the general knowledge on service quality, a greater availability of voltage quality data (on both the transmission and the distribution networks) is a prerequisite for regulators to design focused action plans to improve voltage quality and to introduce regulatory instruments in the interests of customer protection.

As far as measurement at individual connection points is concerned, the situation in Europe is as follows [CEER, 2005].

In most European countries, voltage quality measurements at customers' connection points are carried out upon a customer request. The request can take the form of a complaint, when customers detect problems due to poor voltage quality, or of an enquiry, when customers are simply interested in knowing the voltage quality supplied at their connection point.

In case of a complaint, a large number of countries (though not all) require the distribution company to install a voltage quality recorder at the customer's premises. Generally, the costs of measuring are paid by the requesting customer. In other cases, the costs are paid by the customer if the voltage characteristics comply with the performance standards, and by the company if they do not.

Similarly, in case of a customer enquiry, the network company should be able to provide and install a monitoring system at the customer connection point. Measurement costs are generally paid by the requesting customer. A few European countries allow customers to install their own voltage quality recorder. The voltage quality recorder owned by the customer must comply with technical norms for the measurements to be considered reliable by the distribution company.

In summary, from a regulatory perspective it is important for data on voltage quality to be collected:

- Continuously, at given points on the distribution network, to obtain a general picture of the level of voltage quality (at least for voltage events, since they occur randomly);
- Occasionally, at the point of connection, upon a customer complaint;
- Occasionally or continuously, at the point of connection, upon a customer enquiry.

As of today, regulatory instructions and guidance on data collection are extremely rare, and measurements are typically carried out according to technical norms. Mandatory voltage monitoring at system-level are equally rare. In turn, companies are often required to measure voltage quality at customer connection points when requested to do so by the end-user. A growing number of regulators are engaged in monitoring campaigns in order to measure the actual level of voltage quality before taking regulatory decisions.

## 5.2 Publication

The regulatory instrument of data publication can be employed to inform customers about the general level of voltage quality in the power system, or about the quality supplied in different distribution areas. Obviously, this is only possible if a monitoring system is in place. Measurement systems typically require a test phase, and data should be made public only at the end of the test period.

It may be useful to structure the measured data using aggregated indicators for given supply areas or distribution networks. Detailed, periodic reports can be issued by the regulator or by the network companies. Another effective means of information dissemination is the Internet. For instance, Italian customers have free access to a voltage quality database posted on a dedicated website.[12] The database contains the data collected by the monitoring systems installed on MV distribution networks, and it is updated weekly with actual measures (collected by two sets of voltage quality recorder: 400 recorders are installed on MV-busbars of HV/MV transformers and 200 are installed on MV customer connection points).

Voltage quality data measured at specific points of connection should be made available to the interested customers.

For example, in Portugal, a distributor that has carried out, at the request of a customer, a period of voltage quality measurements, is obliged to communicate to the customer the following information [CEER, 2005]:

[12] The address is http://queen.ricercadisistema.it.

- Monitoring period;
- Type of equipment that was used in the monitoring;
- Type of perturbations that have been registered;
- Analysis of the measure with respect to technical norms;
- Entity responsible for the disturbances;
- Deadline to solve the detected problem if technical norms are not met.

In addition, distribution companies should use system-level data to explain, upon a customer request, how the general level of voltage quality on the network influences the quality supplied at the customer connection point.

For instance, the Norwegian regulation obliges network companies to provide, at the request of current or future customers, information regarding continuity of supply and voltage quality at their point of connection, within one month [Brekke, 2006]. Among other things, companies have to provide the following information:

- The level of nominal supply voltage and the standards for voltage quality indicators;
- Results from fault analyses;
- Number of historic and estimated number of future long interruptions, short interruptions, voltage dips, and voltage swells;
- Calculated minimum and maximum short-circuit power;
- Special conditions in the network that may have an effect on the quality of supply.

Finally, any measured data that are published or communicated should be accompanied by sufficiently specific information regarding traceability, accuracy, calibration interval and measurement methodologies. Technical norms are very useful in this regard and should be employed as a reference.

## 5.3 Minimum quality standards

In the area of voltage quality, the regulatory instrument of minimum quality standards (MQS) is still not employed if we refer strictly to the technical indicators defined in Section 5.1. It is important to note that the limited availability of voltage quality measures at system-level, as well as for the individual customer, is a severe limitation to the introduction of MQS. A second problem, as discussed below, is the absence of enforceable performance standards for several quality indicators.

In Europe today we can distinguish three types of approaches to the problem of ensuring a minimum level of voltage quality to end-users.

In the first and most common case, the regulatory instrument of MQS is used, but only for 'non-technical' aspects of voltage quality. Regulatory intervention is limited to the introduction of MQS requiring that companies respond to customer complaints related to voltage quality within a maximum 'waiting time' and/or carry out individual measurements. As for the more technical aspects, regulators take the technical norms as a reference. However, to date, these technical norms are not used as performance standards in regulatory instruments of the type described in this handbook. As explained below, this approach has a number of drawbacks, and it is no longer considered satisfactory from the point of view of regulators. A joint effort to modify this situation is currently being pursued at the European level.

A small number of European regulators (including, in particular, the Norwegian regulatory authority, NVE) have introduced measuring rules, as well as *voltage quality standards*,[13] on technical indicators of voltage quality. Three points are relevant in this case. First, voltage quality regulation in Norway is the latest development in a regulatory process that started several years ago with continuity of supply regulation [Seljeseth et al., 2005]. Second, the regulatory decisions were taken after years of voltage quality measurements, carried out in cooperation with distribution companies, and they coincided with the introduction of mandatory voltage quality monitoring. Third, even in this case, regulatory intervention is still at a preliminary stage compared to other areas of service quality. NVE has set regulatory instructions on data collection as well as a number of voltage quality standards that differ from the binding limits (see *infra*) indicated in the EN 50160. However, procedures and methodologies to manage violations of voltage quality standards are still to be developed. One important open issue, among others, is the attribution of responsibility for the breach of those standards. This is indeed a rather difficult issue in this area of service quality regulation.

A third approach is the introduction of premium quality contracts. In this case, the existing rules regard measurements, voltage quality standards, and the financial consequences of breaching those standards.

In summary, voltage quality regulation in Europe is just beginning. With good reason, regulators are today mostly engaged in collecting data on the characteristics of the voltage supplied by distribution companies. Indeed, accurate data are the preliminary and crucial prerequisite for the design of any regulatory instrument. As discussed below, another important area of work is the revision of the EN 50160.

In this handbook we describe the 'state of the art'. The first approach is discussed in Section 5.3.1. Voltage quality standards that differ from the EN 50160 are presented in Section 5.3.2, together with more information on the Norwegian regulation. Premium quality contracts are treated in Section 5.4.

### 5.3.1 MQS and the EN 50160

In the majority of European countries, the regulatory instruments introduced in the area of voltage quality concern 'non technical' indicators. For instance, it is common to have GS on investigating voltage complaints, where the regulated indicator is the time between the customer request and the actual provision of the service (the

[13] We use the expression 'voltage quality standards' instead of 'performance standards' because the latter has been used in the rest of this handbook to indicate an element of the regulatory instruments. Unlike performance standards, voltage quality standards are not yet incorporated within a regulatory instrument.

waiting time). A GS of this type is enforced, for example, in Hungary (see Table 3.5 in Section 3.3.2.2).[14] The GS obliges the company, upon receiving a voltage complaint, either to contact the customer, within 10 working days, and offer to adjust the voltage, or to propose a date for installing a quality recorder. Measurements should begin within 5 days, and the company has to inform the customer about the outcomes within 15 days after completing the measures. Furthermore, OS are or have been employed in voltage quality regulation. For example, the UK used to have an OS requiring all voltage complaints to be corrected within 6 months [Ofgem, 2001a].

Another important area of voltage quality regulation concerns voltage disturbances caused by customer equipment. Customers are normally required to minimise the potential disturbances and, in case of specific problems, a number of regulators require distribution companies and customers to find a mutually satisfying solution (within certain time limits). However, defining an acceptable emission level for customers connected to the network is a complex matter, as it involves both the characteristics of the customer installations and the characteristics of the network, and it demands further analysis (within the field of electromagnetic compatibility).

MQS of the type described above constitute a first, important step in approaching voltage quality. However, as with all 'technical' aspects, regulators have so far mostly relied on the EN 50160. It is therefore important to understand why, in recent years, as the attention given to voltage quality has increased, regulators have begun to regard the EN 50160 as not being completely adequate for the purpose of customer protection.

The EN 50160 defines and describes the characteristics of the supply voltage in public LV and MV distribution networks under normal operating conditions. In other words, the voltage characteristics are described as they are found in electricity distribution networks, even in the worst ones [Bollen, 2000].

The distinction between voltage variations and voltage events is very useful for understanding the content of EN 50160. As for voltage variations, the EN 50160 indicates that the voltage characteristics or the relevant indicator "shall be within a range of" or "should not exceed" certain values for a given time interval (*binding limits*). In turn, for voltage events the EN 50160 provides only *indicative values* of the frequency with which the events can be expected to occur.

In terms of voltage variations, the EN 50160 sets the following *binding limits*:

- *Supply voltage variations*. LV: during each period of one week 95% of the 10-minute mean, rms values of the supply voltage must be within the range of ±10% of the nominal voltage $U_n$. MV: ±10% of the declared voltage $U_c$. In addition, for LV only: during each period of one week 100% of the 10-minute means, rms values of the supply voltage must be within the range of +10 %/ -15 % of the nominal voltage.
- *Harmonic voltages*. LV and MV: during each period of one week, 95% of the 10-minute mean rms values of each individual harmonic voltage must be less than or equal to the value given in a table (which, for the sake of brevity, we do

[14] Another GS of this type was described in Section 3.3.2.1, with reference to the UK.

not report here) for harmonics up to the 25th.[15] For instance, 6% for the 5th harmonic, 5% for the 3rd and 7th harmonics, 3.5% for the 11th harmonic, and so forth. Moreover, the THD must be less than or equal to 8%.

- *Supply voltage unbalance*. LV and MV: during each period of one week, 95% of the 10-minute mean rms values of the negative phase sequence component of the supply voltage must be within the range of 0 to 2% of the positive phase sequence component. In some areas and under some circumstances, unbalance up to 3% may occur.
- *Flicker severity*. LV and MV: in any period of one week, the long-term flicker severity caused by voltage fluctuations should be $P_{lt} \leq 1$.
- *Mains signalling voltage*. LV and MV: over 99% of a day, the 3-second mean of signal voltages must be less than or equal to the values given in a figure (which, again, for brevity's sake, we do not report here).[16] For example, in LV networks, 5% of $U_n$ for frequencies between 1 and 10 kHz.

In terms of voltage events, the EN 50160 provides the following *indicative values*:

- *Rapid voltage changes*. LV: a rapid voltage change generally does not exceed 5% $U_n$, but a change of up to 10% $U_n$, with a short duration might occur some times per day in some circumstances.[17] MV: rapid voltage changes generally do not exceed 4% $U_n$, but changes of up to 6% $U_n$, with a short duration might occur some times per day in some circumstances.
- *Supply voltage dips*. LV and MV: the expected number of voltage dips in a year may be from up to a few tens to up to one thousand. The majority of voltage dips have a duration of less than 1 second and a depth of less than 60% of $U_n$ (of $U_c$ in MV).
- *Temporary over-voltages*. (*voltage swells*). LV: under certain circumstances, a short-circuit will produce temporary over-voltages between live conductors and earth. Such over-voltages will generally not exceed 1.5 kV rms.[18]
- *Transient over-voltages*. LV: transient over-voltages between live conductors and earth generally will not exceed 6 kV peak, but higher values occur occasionally MV: no indicative value.

It is evident that, when the EN 50160 provides only indicative values on voltage events, this is not useful for regulatory purposes. The absence of binding limits prevents regulators from adequately protecting customers. For instance, GS cannot be linked to indicative values. In addition, customers are not in a position to claim damages when they occur or to design their own protection systems in an economically sound manner.

[15] Table 1 in the EN 50160 for LV and Table 2 for MV.

[16] Figure 1 in the EN 50160 for LV and Figure 2 for MV.

[17] A supply voltage change resulting in a voltage less than 90% $U_n$ is considered a supply voltage dip.

[18] For the sake of brevity, the corresponding indicative values for MV networks are not reported here.

The only useful values for voltage quality regulation are those listed above as binding limits (for voltage variations). However, even these are not always satisfactory [ERGEG, 2006]. Consider, for instance, supply voltage variations. This voltage disturbance is an essential aspect of distribution service quality because it affects all customers on MV and LV networks. Indeed, rural customer, in particular, can experience today supply voltage variations that exceed the limits imposed by the EN 50160. Nonetheless, the current binding limits on supply voltage variations present the following shortcomings:

- As for LV networks, the more restrictive values ($\pm 10\%$ of the nominal voltage) are binding only for 95% of the time and under normal operating conditions. This means that the EN 50160 exempts network operators from meeting the binding limits in an extremely high number of hours: 8 hours in a week, plus the abnormal situations (these include situations during which the operator carries out maintenance and construction work, as well as exceptional situations outside the operator's control, such as severe weather conditions, natural disasters, acts of public authorities, and so on). Binding limits for 100% of the time are given only over a larger voltage magnitude range (+10%/ -15% $U_n$). In contrast, regulators would be in favour of setting limits for 100% of the time and to evaluate the costs and benefits of a stricter range for voltage variation (for instance, $\pm 7{,}5\%$ $U_n$).
- As for MV networks, the binding limits on voltage variations apply, again, for 95% of the time. In addition, there are no further limits that hold for 100% of the time. The same difficulties are thus envisaged as for LV networks. An additional difficulty for MV customers is the (non-mandatory) use of the declared voltage instead of the nominal voltage. It is possible for the supply voltage at customer connection points to be steadily different from the nominal voltage, without an agreement in this sense with the network operator. In contrast, it would be more useful for MV customers if voltage variations were defined around a 'calculated' supply voltage (computed *ex ante*, for each single point of connection at MV level). Alternatively, it would be even better to have binding limits on the allowed variation of the declared voltage around the nominal value (e.g., $U_c$ within $\pm 5\%$ $U_n$) and stricter binding limits on the supply voltage variations around the declared value (e.g., $\pm 5\%$ $U_n$), for 100% of the time.[19]

It is important to note, however, that defining voltage quality standards that differ from the EN 50160 is not an easy task. For this reason, in addition to the work carried out individually, regulators are also engaged, together with CENELC, in a joint effort to improve the existing technical norms. In December 2006, the European Regulatory Group for Electricity and Gas (ERGEG) published a consultation paper on this subject [ERGEG, 2006]. ERGEG members believe that the EN 50160 should be revised, taking into account the actual levels of voltage quality supplied

[19] Binding limits of this sort are applied in France (see Section 5.3.2).

on European transmission and distribution networks (generally already better than described in the EN 50160), as well as the evolution of customers' needs.[20]

In particular, the ERGEG (2006) consultation paper contains the following recommendations for CENELEC:

1. *Improve definitions and measurement rules*: better definitions and measurement rules are necessary to avoid ambiguities;
2. *Define limits for voltage variations (in particular, for supply voltage variations)*: binding limits given over 95% of time should be upgraded to 100% of the time under normal operating conditions (which already exclude a long list of exceptional situations).
3. *Enlarge the scope of EN 50160 to High Voltage (HV) and Extra High Voltage (EHV) systems*: EN 50160 does not apply to voltage levels above 35 kV; however, in some countries the MV level of distribution networks includes higher voltage levels;
4. *Avoid indicative values for voltage events*: the indicative values in the EN 50160 span wide ranges (for instance: the number of voltage dips is between a few tens and one thousand per year, and the number of short interruptions is between a few tens and several hundreds per year). A reason for this is that no distinction is made between events with different characteristics. A preliminary step to the introduction of binding limits would be, for instance, to classify voltage events according to severity;
5. *Consider duties and rights for all parties involved*: a separation of responsibilities between network operators and equipment manufacturers (and customers) would be highly beneficial. This requires coordination with the existing technical norms for customer appliances;
6. *Introduce limits for voltage events, differentiated according to the network characteristics*: binding limits should be set for those voltage events that are both severe and under the responsibility of the network operator. Such limits should be differentiated according to the characteristics of the network.
7. *Premium quality contracts*: the introduction of guidelines on premium quality contracts is deemed extremely useful to help customers and network operators to employ these instruments.

In other words, regulators envisage gradual changes, that begin with improved measurement rules and definitions (point 1). In particular, as better explained in Section 5.5, two essential steps in the direction of introducing standards on voltage events are, then, a classification of the severity of the events (point 4) and the identification of responsibilities of all interested parties (point 5). In this regard, a balance is necessary between the quality supplied by the distribution network and the sensitivity of customer appliances. Standards are necessary only for those events that actually cause malfunctioning or damage to the end-user equipment (i.e., are severe) and that can reasonably be assigned under the responsibility of the

[20] European energy regulators first expressed their concerns with respect to the EN 50160 in the Third Benchmarking Report on Quality of Electricity Supply [CEER, 2005].

network operator (point 6). As it is done for continuity of supply, also voltage quality standards must take into account differences in network technical characteristics.

Regulators are also aware of the fact that the decision regarding voltage quality standards might be delicate. If these standards are set close to the actual values of the voltage characteristics, therefore, they might not be excessively challenging for network companies; their main objective would be to prevent deterioration of voltage quality levels. If standards are set at a more demanding level, this choice might also entail higher investment costs for network companies, depending on the actual level of voltage quality already supplied. In this latter case, regulators should allow tariffs to increase in order to assure necessary investments.

After the publication of the ERGEG consultation paper, the Council of European Energy Regulators (CEER) has begun cooperating with CENELEC to review the EN 50160 with the objective of introducing binding and more realistic limits for voltage disturbances.[21] The availability of field data from the monitoring systems recently installed in several countries will be extremely helpful for this purpose.

### 5.3.2 Beyond the EN 50160

A number of European countries have introduced voltage quality standards that tighten the requirements contained in the EN 50160 (France, Hungary, Norway, and Portugal, and Spain), although only in Norway did this happen at the initiative of the regulator. Various voltage disturbances have been addressed, including:

- Supply voltage variations;
- Rapid voltage changes;
- Flicker severity;
- Voltage dips;
- Transient over-voltages;
- Voltage unbalances;
- Harmonic voltage distortions.

For details on these voltage quality standards we refer the interested reader to CEER (2005). Nonetheless, it is interesting to report here the voltage quality standards defined for supply voltage variations. Indeed, this disturbance is of primary concern for regulators. CEER (2005) reports that the following standards apply:

- *France*. On MV networks, the supply voltage must be in the range of $\pm 5\%\ U_c$ for 100% of the time, where $U_c$ must be in the range of $\pm 5\%\ U_n$;
- *Hungary*. On LV and MV networks, the supply voltage must be in the range of $\pm 7{,}5\%\ U_n$, with a maximum of 115% for not longer than 1 minute;
- *Portugal*. On EHV and HV networks, during each period of one week, 95% of the 10-minute mean rms values of the supply voltage must be within the range of $\pm 5\%\ U_c$; $U_c$ must be in the range of $\pm 7\%\ U_n$;

[21] A revised version of the EN 50160 will be published in 2007, but with only minor modifications. If accepted, the revisions indicated in ERGEG (2006) will be included in a further revision of the technical norm.

- *Norway*. On LV networks, 100% of the 1-minute mean rms values of the supply voltage must be in the range of ±10% $U_n$;
- *Spain*. The supply voltage must be in the range of ±7% $U_c$.

In the past few years, the Norwegian regulator has dedicated considerable resources to the area of voltage quality regulation. NVE has been working closely with distribution companies in the collection of data on voltage characteristics and in publication of the relevant statistics, and it has provided constant assistance to customers and network companies in order to solve specific voltage quality problems. It has also conducted surveys on the costs incurred by customers for power quality, and has carried out a broad consultation process on the issue [Brekke, 2006].

There are a number of aspects in the most recent regulatory decision that are relevant in the context of setting voltage quality standards [NVE, 2005]. The Norwegian regulation defines both measurement standards and values of the voltage characteristics. Based on the practical experience gained in recent years, NVE has set standards which in a number of cases are more demanding than the values given in the EN 50160. A comparison is given below (the corresponding values indicated in the EN 50160 are given in parentheses) [Brekke, 2006, Seljeseth et al., 2005].

- *evaluation period of voltage disturbances*: values for voltage unbalances, harmonic voltages and flickers: 100% of the time (95% of the time);
- *rms time interval*: supply voltage variations are measured using a 1-minute mean rms values (10-minute mean);
- *flicker severity*: $P_{lt} \leq 1$ for 100% of the time and $P_{st} \leq 1$ for 95% of the time ($P_{lt} \leq 1$ for 95% of the time);
- *rapid voltage changes* (LV network): under normal operating conditions, a rapid voltage change of up to 10% $U_n$ can only occur, at the most, once per day, and a rapid voltage change of up to 5% $U_n$ can occur a maximum of 24 times per day; rapid voltage changes occurring more than 24 times per day must not exceed 3% $U_n$ (under normal operating conditions, a rapid voltage change generally does exceed 5% $U_n$, but a change of up to 10% $U_n$ might occur some times per day in certain circumstances);
- *harmonic voltage distortion*: the THD of the supply voltage must be less than or equal to 8%, and the THD weekly average must be less than or equal to 5% (the THD of the supply voltage must be less than or equal to 8%);
- *higher harmonics*: maximum values are given, but for brevity's sake we do not indicate them here (no values are given for harmonics of order higher than 25, as they are usually small, but largely unpredictable due to resonance effects);
- *EHV and HV networks*: values are given on rapid voltage changes, flickers, harmonics, voltage unbalances (no values are given).

It is also important to note that, for certain voltage disturbances, NVE has chosen not to introduce standards. In particular, these disturbances are:

- Voltage dips;
- Short and long interruptions;

- Temporary over-voltages (swells);
- Transient over-voltages;
- Interharmonic voltages;
- Mains signalling voltage.

In this regard, the decision not to introduce standards is due to the difficulties in defining binding values and monitoring them, or to the small socio-economic importance of some of these disturbances [Seljeseth et al., 2005].

The Norwegian regulator recognizes that the recent decision raises a significant number of issues, opening the field to further research in both technical and economic areas. In the technical field, methods and concepts for the analysis and optimization of distribution networks are needed which account for voltage quality standards, together with systems for measurements and collection of large amounts of voltage quality data. In the economic area, procedures and methodologies are needed to manage violations of the regulatory standards and to identify the party responsible for the problem. Finally, a higher coordination and harmonization is also necessary between national regulators and international bodies responsible for technical standardization [Seljeseth et al., 2005].

In summary, we can certainly expect that voltage quality will draw increasing attention on the part of regulators in the coming years. In particular, the issue of the regulatory inadequacy of EN 50160 makes the definition of voltage quality standards, together with measuring campaigns, a priority in several countries.

## 5.4 Premium quality contracts

Premium quality contracts are an interesting option for voltage quality regulation. Although not widely employed, they are regarded as an efficient solution for improving quality without adding extra costs to the general tariff. The basic features, as well as the main advantages and disadvantages of these contracts, were presented in Section 4.5. In this section we focus on practical aspects.

A common type of premium quality contract is one that defines specific conditions for connection to the grid. The customer can obtain a higher level of continuity of supply and voltage quality, for instance, by requesting a dual connection with an automatic changeover.

Other types of contracts are available in France, Italy, and Norway.

In France, both transmission and distribution companies offer customers the possibility to contract for higher levels of quality using a customized contract. As far as voltage quality is concerned, the standard contract for MV customers obliges the distribution company to keep supply voltage variations, flickers and supply voltage unbalances within values provided for in the contract [ERD, 2000]. A customized contract also covers voltage dips.

For customers connected to the transmission network (63 kV and above), the maximum number of voltage dips (resulting in a voltage of less than 70% $U_c$ and longer than 0.6 ms) is 5. Voltage dips occurring less than one second after an interruption (short or long) are not counted, nor are those due to a fault in the customer's installation. Other exemptions apply for customers connected at 225 kV and 400 kV. For customers connected to the MV distribution network, the maximum number of voltage dips (resulting in voltage of less than 70% $U_c$ and longer than 0.6 ms) depends on the point of connection; however, it cannot be less than 5. Customized contracts are not available for LV customers. Customized contracts imply the payment of an annual fee to the network operator and make the latter responsible for all direct and certain damages caused to the customer.

The use of customized contracts, a practice dating back to before the regulatory authority was constituted, is rather frequent. For obvious reasons, customers with a customized contract must have a voltage quality recorder (the owner of which can be either the customer or the network operator). The percentage of customers with quality recorders installed at their connection points is significant: 16% on MV distribution networks, and 12% on the EHV and HV transmission network.

In Norway, customers and distribution companies are allowed to enter into individual contracts where they agree on either lower or higher power quality levels than indicated in the regulatory framework. Distribution companies are not allowed to enter collective contracts with several customers. However, the level of quality can be adjusted for a group of customers if each customer in the group has signed an individual contract [Seljeseth et al., 2005].

The Italian contracts were described in Section 4.5.

## 5.5 Advanced topics

The regulation of voltage quality is a complex matter and the issues that require further analysis are numerous. We focus here on a well-defined problem - the definition of binding values for voltage dips - in order to illustrate the difficulties that regulators face when working in this area.

As briefly mentioned in Section 5.3.1, two essential steps are necessary to arrive at a definition of binding values on voltage magnitude events (in particular, voltage dips and swells): (*i*) the identification of responsibilities of all interested parties: network companies, customers and equipment manufacturers, and (*ii*) a regulatory classification of the severity of the events.

As far as responsibility is concerned, it is fundamental to observe that the sensitivity of end-user electrical equipment varies according to the depth (or height) and duration of the voltage events. End-user equipment will generally continue to operate during the occurrence of voltage events of specific duration and amplitude, but it will stop functioning for events with different characteristics. The so-called 'voltage tolerance' of equipment is conventionally illustrated in the ITI Curve, depicted in Figure 5.9.[22] The ITI Curve describes the voltage which can typically

[22] This curve accurately reflects the performance of typical single-phase, 60-Hz computers and their peripherals, and other information technology items such as copiers, fax machines, and point-of-sales terminals. While specifically applicable to computer-type

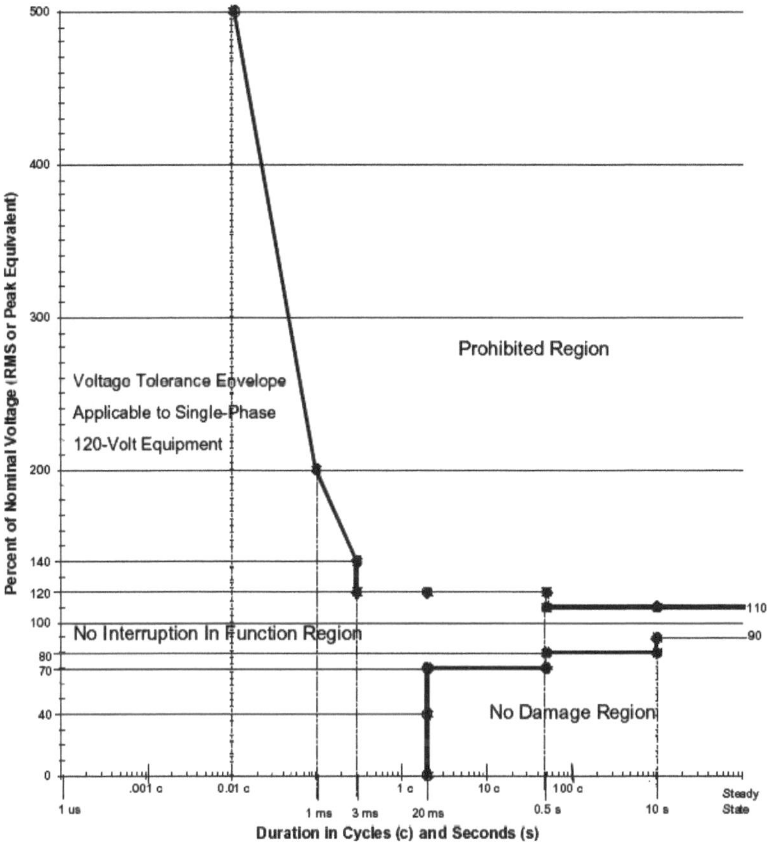


**Fig. 5.9.** The Information Technology Industry Curve [ITIC, 2000].

be tolerated ('no interruption in function' region) by most Information Technology Equipment (ITE). The 'no damage region' indicates that the normal functional state of the ITE is not typically expected during these conditions, but no damage to the ITE should result. The 'prohibited region' indicates that, if ITE is subjected to these conditions, damage to the ITE may result [ITIC, 2000].[23]

From a regulatory perspective two points are important. First, the effort of reducing voltage dips (and swells) can be limited to the region where the end-user equipment is indeed sensitive. Second, the sensitivity or tolerance of the equipment also depends on the manufacturing industry, or on the choice of the equipment made by the customer. Thus, when working on defining binding values

---

equipment, the ITI Curve is generally applicable to other equipment containing solid-state devices.

[23] The ITI Curve is clearly not intended to reflect the performance of all end-user electronic-based equipment. The issue of voltage tolerance curves for different categories of end-user equipment is beyond the scope of this handbook.

on voltage quality indicators, balance between technical requirements and costs should be considered. Reducing voltage disturbances only through interventions on the networks could be inefficient (and costly). Electrical equipment manufacturers and customers can contribute to the effort of mitigating the effects of voltage disturbances.

The sharing of responsibilities between all the parties involved is clearly important for the development of an efficient and effective regulatory framework. An international working group of regulators recently proposed to integrate the EN 50160 with the indication of two 'regions of responsibilities': an area, in the voltage-time space, where regulation is not necessary (the end-user equipment is able to operate satisfactorily) and an area where the frequency of the events will be regulated (with instruments to be defined) [Bollen and Verde, 2007, ERGEG, 2006].

As discussed in Section 5.1.1, the frequency of voltage dips is normally measured with a table indicating ranges of magnitude and duration (see Table 5.1). Thus, setting binding values for the frequency of these events implies a definition of the relevant magnitude and duration ranges, as well as a definition of the maximum number of dips within each range. The former is often referred to as a classification of voltage dips according to their severity. To date, an international consensus on this classification has not been reached and a European classification table does not exist.

Defining groups of events that present similar characteristics of severity is rather complex. A classification table should take into account several aspects:

- Electrical equipment tolerance curves and different potential consequences (trip, malfunction, damage, and so forth);

<table>
<tr><th rowspan="2">Residual voltage U (as % of declared voltage)</th><th colspan="3">Duration $t$</th></tr>
<tr><th>$20 \leq t < 150$ ms</th><th>$150 \leq t < 600$ ms</th><th>$0,6 \leq t < 3$ s</th></tr>
<tr><td>$90 > U \geq 85$</td><td colspan="3" rowspan="1">Y*</td></tr>
<tr><td>$85 > U \geq 80$</td><td colspan="2"></td><td rowspan="2">Z1</td></tr>
<tr><td>$80 > U \geq 70$</td><td></td><td rowspan="3">S1</td></tr>
<tr><td>$70 > U \geq 60$</td><td>X1</td><td rowspan="3">Z2</td></tr>
<tr><td>$60 > U \geq 40$</td><td>X2</td></tr>
<tr><td>$40 > U \geq 0$</td><td colspan="2">T1</td></tr>
</table>

*minimum plant compatibility requirement

**Fig. 5.10.** Classification of voltage dips [NRS, 2004].

- Different typical causes for the voltage event (short-circuits, earth faults, start-ups of motors, etc.);
- Location of the fault (close or remote).[24]

An interesting classification is adopted in South Africa and is reported in Figure 5.10 as an example [NRS, 2004]. This table was built accounting for potential consequences of the events, and typical causes (it is not a regulatory document but a technical norm).

Note that voltage dips are classified by residual voltage and duration in seven categories (Y, X1, X2, T1, S1, Z1, and Z2). In particular, they include those events which are below 90% of the declared voltage and which have a duration between 20 ms and 3 s. The frequency of voltage dips is measured as the number of events within each category over a given period of time.

Based on this classification, customer installations are expected to tolerate the events in category Y, as they are expected to occur frequently. One of the regulator's priorities should be to introduce binding limits on the frequency of the events in categories X1 and X2, i.e., voltage dips that might effect sensitive customer equipment.

[24] The location influences the depth of the voltage event.

# 6

# References on specific issues

This chapter collects specific information on three practical aspects of a regulator's activities which have been mentioned throughout the handbook but which have not yet been fully examined: the consultation process, customer surveys, and auditing procedures. These complementary tasks are essential to the main activity of designing a regulatory framework for service quality. They fundamentally support the decision-making process, and they are crucial components of everyday practice at the implementation stage. In this chapter, without meaning to be exhaustive, we give some indications as to how consultation, surveys and audits may be addressed. In particular, the consultation process is discussed in Section 6.1, details on customers surveys are given in Section 6.2, and auditing procedures are described in Section 6.3.

## 6.1 The consultation process

Consultation is a crucial aspect of the decision-making process of a regulatory authority. An open and thorough consultation improves the quality of the decisions and, by increasing transparency and understanding of the regulator's work, it creates consent with respect to the policy solutions that are devised, proposed and deployed.

Usually, each regulatory authority delineates its own consultation policy, defining objectives and procedures of the process. One of the most advanced consultation processes is implemented in the UK. The details are described in Ofgem (2002b). Below we briefly present the main characteristics of a consultation process, discuss a few practical aspects, and make some final remarks.

### 6.1.1 The process

A consultation process consists of four basic steps: the publication of a consultation document, the collection of responses, the holding of informal meetings with interested parties and formal hearings with the regulatory authority commissioners, and

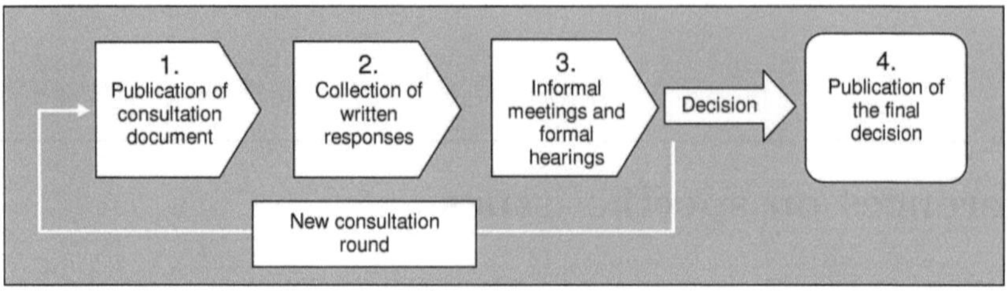


**Fig. 6.1.** The consultation process

the issue of the final decision. However, as illustrated in Figure 6.1, it is not unusual to go through the first three steps a few times (consultation rounds) before a final decision is made.

When the regulatory authority starts working on a particular subject, it publishes a consultation document setting out the issues, describing the possible solutions, and inviting views. The first consultation document is usually devoted to defining the general principles and objectives of the policy. A number of possible options are proposed for consideration, all complying with the general principles and objectives.

The range of these options can be narrow or broad, depending on the issue being examined. When the issue is particularly complex or was never addressed before, the alternatives can be numerous. These different solutions generally present different levels of regulatory control over the market. They may range from 'doing nothing' (i.e., simply removing the existing barriers for the market to work efficiently), to introducing mandatory requirements for the companies to follow. By contrast, when the issue under examination is simpler, or it has already been addressed before, the alternatives presented in the consultation document focus more on the details of the regulatory policy and/or on implementation problems.

In the most advance cases, a study of the anticipated effects of the main policy options is included. This analysis is called a 'Regulatory Impact Assessment' (RIA). A RIA considers the impact (i.e., the benefits and costs) of some of the proposed solutions (normally at least a couple of them) for those directly affected by it, including the regulated companies and the customers, and its indirect, potential consequences for the environment, the level of competition in the market, and so on.[1]

Consultation documents state clearly all open issues, and they contain data as well as quantitative analyses in order to create an objective framework for discussion. A regulatory policy affects numerous parties, with different backgrounds and expertise. When the consultation process forces them to interact, the regulator should make an effort to create a level playing field and to ensure that a fair debate takes place, notwithstanding the asymmetry of information between the parties.

The language used in a final decision is usually legal and/or technical; similarly, the structure of the decision document is fixed by law. By contrast, in consultation

[1] Guidance on conducting impact assessment can be found in Ofgem (2005c).

documents, regulators use simple and plain language so that all interested parties can understand them. In addition, these documents do not have the structure of a final decision because this would prevent the regulator from proposing different alternatives.

Although every consultation involves the publication of the relevant documents, the regulator may also use seminars to add to the effectiveness of the consultation. In addition, seminars also enable the less informed stakeholders (such as consumer associations, environmental groups, or labour unions) to acquire a better understanding of the issues and thus to provide their contribution to the discussion.

Once responses and contributions have been collected, the regulatory authority departments may organize meetings with respondents. These meetings are particularly important because they enable the regulatory authority to receive clarifications on the respondents' point of view, to analyse their proposals further, and to discuss technical details.

Seminars and meetings are normally attended by the departments of the regulatory authority. They may be followed by one or more formal hearings, conducted by the authority commissioners who are entitled to make the final decisions. Hearings are public and open to debate and discussion, although (understandably) they are mainly devoted to the general principles of the policy under approval.

If the issue is not complex and the comments are broadly supportive, public hearings are immediately followed by a final decision, published in the prescribed legal form.

If, by contrast, the extent and complexity of the issue raised by the consultation require an additional consultation round, a second consultation document is issued. This document will take into account the responses already received and, just as before, it will invite views on a number of alternative solutions. In the simplest cases, this second consultation document may already contain a draft of the final decision. Normally this means that the only questions that remain open regard technical and economic parameters. If this is the case, the consultation document will typically propose a range of values for those parameters to simplify the process.

For more complex issues, the second consultation document will deal again mostly with principles and objectives, and the consultation process will continue for as many rounds as necessary.[2]

### 6.1.2 Practical aspects

A few practical aspects of the consultation process are highlighted below.

First, consultation documents are usually published on the website of the regulatory authority. For major policy initiatives, the regulator may also write to interested parties and enclose copies of relevant documents.

Second, the time allowed for submitting responses (i.e., the consultation period) depends on the nature of the issue under consideration and/or on the timing of the

[2] For complex issues, in the UK, it is not unusual to go through 6 or 7 rounds.

consultation itself. Normally it is not less than four weeks, and in any case the time period should be indicated clearly for each consultation round. More time may be allowed for responses if, for example, the policy proposal is particularly complex, or if the policy is at a very early stage of development.

Third, the regulator usually accepts all type of written responses (e-mails, faxes, letters) and, in some cases, the regulator sends a letter acknowledging receipt of the response. The regulator does not reply in detail to every submission. Nonetheless, the key issues raised by respondents are dealt with in the next document published in the series. Once the consultation period is closed, responses can be placed on the regulator's website. Publishing responses helps to stimulate debate.

Following consultation, the time taken for the regulator to react - with either a further consultation document or a decision document - depends on the issues raised by the consultation and on the prioritization of its resources.

### 6.1.3 Remarks

The main function of the consultation process is to improve the quality of the final decision. It is therefore essential for the consultation process to be effective.

The effectiveness of a consultation process can be verified by comparing the initial proposal with the final decision. Only when all the relevant issues raised by the respondents have been adequately considered in the final decision has the function of the consultation process been fulfilled. In practice, consultation and decision documents often include a 'motivation' section where the regulator collects and comments on the responses, and explains why, in view of the opinions received, it is making a particular decision.

Contributions from interested parties usually deal with practical implementation issues and they serve to alert the regulator as to the feasibility of proposed measures. In particular, it is extremely important to take account of contributions from the companies that are to be subject to the proposed measures if a workable final decision is to be reached. Indeed, both feasibility and consent are highly desirable features of any regulatory decision. They provide incentives for companies to actually comply with the decision (and not to circumvent it), and they help to avoid litigation before the courts. This is why the potentially substantial administrative work occasioned by the consultation process should be regarded not as a burden but as an investment.

## 6.2 Customer surveys

When designing service quality regulation, results from customer surveys can make significant contributions to the decision-making process. In particular, regulators have used customer surveys for three main reasons:

- to acquire information on customer *satisfaction* with the service provided by the electricity distribution companies;

- to identify customer *priorities* and *expectations* for improvements in quality of service;
- to estimate customer *willingness to pay* (WTP) for improvements in service quality.

In this section, we illustrate how the information listed above can be collected using different types of surveys. Clearly, the content of the questionnaire, the survey procedure, and the analysis of the results depend on the main objective of the survey. In that regard, regulators also carry out surveys with objectives other than those listed above: for instance, there are the surveys concerning the quality and speed of telephone responses described in Section 3.4 (e.g., the call-back survey). Nonetheless, the focus of this section is on surveys that can provide guidance in the design of the regulatory scheme. In particular, Section 6.2.1 focuses on customer satisfaction, priorities and expectations. Section 6.2.2 addresses customer interruption costs (i.e., WTP for improvements in continuity of supply).

### 6.2.1 Customer satisfaction, priorities and expectations

Customer *satisfaction* with the level of service quality offered by the distribution companies is the simplest information that can be obtained with a survey, and at the lowest cost. Surveys of this sort are often carried out by regulators on a regular basis (e.g., once a year).

For this type of survey, it is not uncommon to rely on the national body responsible for official statistics (or similar agencies). The customer satisfaction survey can thus be incorporated in a broader survey (hopefully with some logical connection) that is normally administered by that body.

When a so-called 'omnibus research' method is chosen, the customer target (normally households), the sample size and stratification (fundamental for the results to be representative of the customer target), as well as the method of the interview (mail, telephone or face-to-face) are already defined. The regulator only has to identify the issues to be explored. To this end, it is important that the criteria followed in the omnibus research are clear to the regulator, so that an appropriate questionnaire can be prepared and the details of the *ex post* analysis can be defined. As indicated, the set of questions that are relevant for the regulator will be included in a field study with a broader scope. Therefore, it might be convenient to explore satisfaction as regards multiple regulated services (e.g., electricity and gas service quality). The preparatory work for omnibus research is quite limited in time and the costs are modest. In addition, if the study is carried out by the national body responsible for official statistics, a call for tender may not be necessary to assign the job.

A questionnaire on customer satisfaction will address both general and more specific issues concerning the level of satisfaction with commercial quality aspects and with technical aspects of quality. A short list of possible questions is given in Table 6.1. Answers are typically scaled from 'very dissatisfied' to 'very satisfied', or similar scales.

**Table 6.1.** Questions on customer satisfaction - an example

| General questions | |
|---|---|
| General satisfaction with the electricity service | From very dissatisfied to very satisfied |
| Service improvement in the last year | Yes, no, the same, do not know |
| ***Commercial quality aspects*** | |
| Satisfaction with: | |
| Frequency of meter readings | |
| Readability and clearness of bills | From very dissatisfied to very satisfied |
| Convenience in the form of payment | |
| Punctuality in receiving the bills | |
| Satisfaction with: | |
| Customer centre opening hours | |
| Waiting time at the customer centre | From very dissatisfied to very satisfied |
| Waiting time for receiving responses to requests | |
| Quality of information received by telephone | |
| Speed of telephone responses | |
| ***Technical quality aspects*** | |
| Experience (and inconvenience) with: | |
| Planned interruptions | Never, sometimes, frequently, do not know |
| Unplanned interruptions | (From not at all to very inconvenient) |
| Voltage problems | |

However, when carrying out a customer survey, it is important to address additional issues besides satisfaction. In particular, a survey can provide support in identifying which aspects of service are valued by customers and the relative *priorities* placed on different outputs. Another important piece of information that can be solicited from customers is their *expectation* as to the level of service quality. This information can provide guidance in setting performance standards on various quality indicators.

Note that this type of information might be too specific to be collected with an omnibus research approach, and an *ad hoc* survey is generally preferred. In the latter case, costs and efforts on the part of the regulator are certainly higher, but there is also a greater degree of freedom in designing the survey. The survey is normally committed to an independent research company, selected according to the criteria required by law (often by way of a call for tender).

The regulator defines the customer target (often both domestic and non-domestic customers) and, together with the appointed company, makes sure that the sampling size and stratification meet the requirements: geographical coverage (rural and urban customers) and location (different regions of the country), social and economic grouping (for domestic customers), and company size (for non-domestic customers). A decision is made regarding the method of the interview (by mail, phone, face-to-face) with a view toward achieving a satisfactory response rate and robustness in the results. The focus of the questionnaire and its main topics depend primarily on the objectives of the regulator. A number of focus groups and interviews can also be carried out to seek additional opinions on the draft questionnaire from interested parties. In addition, prior to the main fieldwork stage, a small number of pilot interviews are normally conducted to ensure the clarity, flow and understanding of the questionnaire.

Surveys on customer satisfaction, priorities and expectations can be structured in different ways. Here we briefly describe two different approaches. First, we describe the experience of Hungary, where surveys on customer satisfaction and priorities are conducted once a year. Then we discuss an example from the UK, where a survey on customer satisfaction, priorities and expectations was conducted in 2003 during a price control review.[3]

Indeed, it is extremely useful for a regulator to have a better understanding of customer expectations of average levels of quality of service, their relative priorities among potential improvements in different aspects of quality, and their willingness to pay for improvements. Nonetheless, carrying out the survey is only an intermediate step in the decision-making process. How this information is used to reach a final decision is left to regulator's judgement and experience.

#### 6.2.1.1 Hungary

The objective of the annual Hungarian survey is to gather information on customer satisfaction and priorities. The structure of the survey is rather straightforward. Customers are requested to provide an indication of 'satisfaction' on a number of aspects of the electricity service, as well as an indication of their relative 'importance'. The issues investigated with the survey include several commercial quality aspects such as metering and billing activities, responses to complaints, and quality and content of information provided in customer centres. Opinions are also sought on tariffs and technical aspects of quality.

The analysis of the results produces a 'satisfaction' score and an 'importance' score, in each case as a percentage (where 100% indicates high satisfaction and high importance), for each item in the questionnaire, per company and at national level. In addition, a 'satisfaction minus importance' score is derived, also as a percentage, by subtracting the importance score from the satisfaction score.

Data are compared across companies and changes over time are observed. A list of customer priorities is also defined, based on the 'importance' score. Finally, a positive 'satisfaction minus importance' score is interpreted as over-performance: the service offered by the companies is said to meet customer expectations. By contrast, a negative 'satisfaction minus importance' score indicates that company performance falls short of customer expectations [Rekettye and Orosdy, 2003].

[3] We refer to the first phase of the study conducted by Accent Marketing and Research on behalf of the UK regulator [Ofgem, 2003].

#### 6.2.1.2 United Kingdom

An important characteristic of the UK study was that the survey was part of the distribution price control review. The UK regulator was interested in reviewing the existing quality regulation and in collecting useful information for the design of a new regulatory scheme. Accordingly, the survey was designed to explore quite a large number of issues, but the questions asked were much more specific than the questions normally included in a periodic or omnibus survey. In fact, a primary objective of the study was to identify priorities for services to be regulated and customer expectations as to performance standards and willingness to pay for improvements.[4]

Instead of describing the general structure of the survey, we briefly focus here on two issues in order to illustrate how information on expectations and priorities can be solicited from respondents. For the full questionnaire and analysis of the results, we refer to Ofgem (2003).

**Guaranteed Standards**

A series of questions were asked to evaluate customers' attitudes toward the guaranteed standards of performance (GS), and to explore how these could be improved. First, the interviewer read out a brief description of the GS. For example [Ofgem, 2003]:

- *Distributors should restore domestic customers' supplies within 18 hours following unplanned interruptions. Failure to do so results in a penalty payment of £50 for domestic consumers for the first 18 hours plus £25 for each additional 12 hours.*

Then, the respondent was asked whether the 18-hour timeframe was 'about right', or whether it was 'too long'. When the response was 'too long', respondents were asked what they thought would be more appropriate.

Respondents were also asked whether they felt that a compensation payment of £50 for a period of 18 hours was considered 'about right', 'too little' or 'too much'. The respondent was also asked to indicate, if relevant, a more appropriate compensation.

Finally, the respondent's attitude towards automatic payment of compensation was examined, together with his or her willingness to pay in order to facilitate investment in technology that would make automatic payments possible.

**Interruptions**

A number of questions in the UK survey explored the respondents' experience with interruptions and the level of inconvenience they caused. The respondents were also asked for their opinion regarding [Ofgem, 2003]:

[4] The willingness to pay results obtained with this study were only indicative; a 'stated preference' research was subsequently undertaken in the second phase of the research (see Section 6.2.2).

- the number of unplanned interruptions that distribution companies should be allowed to have before compensation is paid (on a scale of 1 to 7);
- the expected restoration time after an unplanned interruption (on a scale of 1 to 18 hours);
- the expected restoration time after an unplanned interruption caused by a major storm (on a scale of 1 to 72 hours);
- whether companies should be doing more to reduce the impact of severe weather and what they should do.

In addition, among other things, respondents were asked to rate the following aspects of service (from 'not important' to 'very important') [Ofgem, 2003]:

- Dedicated help-line for those with special needs
- Getting accurate information on when power will be restored
- Rapid restoration of power
- Being able to get through to someone quickly
- Getting useful information
- Receiving compensation

As can be expected, this type of survey provides very specific results. For example, the survey determined, among other things, that [Ofgem, 2003]:

- most customers believe that up to 3 unplanned interruptions lasting more than 3 minutes should be allowed before compensation is paid;
- the majority expect power to be restored 1-3 hours after an unplanned outage, although a longer period is acceptable (for many customers, 24 hours or more) in severe weather;
- most customers felt that penalty payments should be automatic. There was some willingness to pay for this service: 37% percent of the respondents were willing to pay extra. The average amount they were prepared to pay was £2.63, or 3.3% of the average monthly bill.

### 6.2.2 Customer interruption costs

Electricity companies have analyzed customer interruption costs since the 1980s. The primary application of these studies was in cost/benefit analyses related to system planning and expansion, and in the adoption of operating policies. The underlying idea was that customer interruption costs are a good measure of what was called 'reliability worth', or the benefit to customers of uninterrupted supply. Therefore, reliability worth, captured by the use of customer interruption costs, was weighted against the costs of building new generation or transmission capacity that would increase system reliability. Similarly, when considering various alternatives for transmission and distribution network planning, the cost of providing service at different levels of reliability (as a function of system configuration) was weighted against the reliability worth. Finally, customer interruption costs were also used as a criterion for devising maintenance policies for generation plants and transmission networks, as well as load shedding or load restoration sequences. In other words, customer interruption costs were included, together with the costs of the company for reliability-related expenditures, in a total cost function: minimizing this function, an optimal level of reliability was deducted and taken as a reference in the decision-making process [Wacker and Billinton, 1989].

The studies carried out by electric companies have greatly contributed to the an understanding of the nature and variety of customer costs, and of the cost valuation methodologies. Following the literature on the subject, we can briefly introduce a few basic concepts [Allan and Billinton, 2000, Wacker and Billinton, 1989].

First, the impact of an interruption can be classified according to two main categories: short-term and long-term. Short-term impact results directly or indirectly from the supply interruption, while long-term impact results from a response to the occurrence of interruptions.

As far as *short-term* costs are concerned, examples of direct individual impacts include lost production, costs of restarting an industrial process, and equipment damage for a non-domestic customer and spoiled food in the freezer for a domestic customer; direct social impacts include, for instance, inconvenience due to lack of transportation. Examples of indirect individual costs are the damaged reputation of an industrial customer who cannot meet its contractual obligations with clients, or the loss of leisure time (e.g., missing a television programme) for a domestic customer. Indirect social costs derive, for instance, from looting during an extended blackout.

*Long-term* costs include all mitigating actions undertaken by customers to reduce or avoid the potential interruption costs. Examples of such costs range from the installation of back-up generators to the purchase of flashlights and candles.

Second, a number of approaches have been devised to evaluate interruption costs. They can be broadly categorized according to three categories.

- *Indirect analytical methods* which infer interruption costs from associated variables, or proxies. For example, the value of loss production is determined taking the ratio of the annual gross national product to total electricity consumption;
- *After-the-fact case study* of a particular outage. This approach has been applied to the analysis of major, large-scale blackouts in the attempt to capture both direct and indirect short-term costs;
- *Customer surveys.* With this approach, customers may be asked to estimate their direct, short-term costs due to supply interruptions of varying duration, and at different times of the day and year. Alternatively, they can be asked what they would be willing to pay to avoid having the interruption, or, conversely, what amount they would be willing to accept (WTA) for having to experience the outage (*contingent valuation* analyses). Indeed, it is common (and desirable), in practice, to evaluate all three figures: direct costs, WTP and WTA. Indirect valuation methodologies can also be used. For example, customers can be asked what preparatory actions they might take in the event of recurring interruptions (*preparatory action* approach); otherwise, customers can be requested to rank, in order of preference, different and mutually exclusive combinations of prices

and service quality levels (*conjoint analyses*). By using this latter approach, the WTP for quality increases and the WTA for quality degradation can be identified.

Third, the literature indicates that, from a customer's perspective, the cost of an interruption is related to a number of factors. In particular, interruption costs are a function of both the customer characteristics and the characteristics of the interruption.

- *Customer characteristics* include the type of customer (e.g., domestic, non-domestic), the nature of the customer's activities (e.g., industrial, commercial, or essential services, such as hospitals), size of the electrical load, and energy dependency as a function of the time of day and time of the year.
- *Interruption characteristics* mainly include its duration and frequency, and whether the interruption was notified in advance.

In recent years, regulators have also become interested in customer interruption costs (or customer willingness to pay for improvements in continuity of supply). As discussed in Section 4.4.1, monetary valuations for higher or lower levels of quality are essential in the design of reward and penalty schemes. To date, analyses of customer interruption costs undertaken for regulatory purposes (by regulators and/or research institutions) have favoured the customer survey approach, in the form of contingent valuation or conjoint analysis.

The strength of the customer survey approach lies in the fact that customers are probably in the best position to assess their own interruption costs. Direct costs are relatively easy to evaluate for certain customer categories, namely, industrial and commercial customers. On the other hand, direct costs are more difficult to quantify for domestic customers. In order to capture inconveniences, which do not have a direct monetary effect, electric companies have made considerable use of the 'preparatory action' approach. By contrast, the impact of the interruption has often been valued by regulators in terms of WTP.

Contingent valuation analyses have been conducted in Norway, Sweden, and Italy.[5] In terms of content, procedures, and transformation of the surveyed data, regulators have drawn significantly from the framework established by the above-cited literature. However, the application of these studies is obviously different. Customer interruption costs and/or WTP are in fact used to set incentive rates for reward and penalty schemes. The transformation of the surveyed data thus follows the framework indicated in the literature only to a certain extent.

A WTP customer survey was also conducted in the UK. In this case, a conjoint analysis approach was preferred.[6]

---

[5] For Norway, see Samdal et al. (2003); for Sweden, see Carlsson and Martinsson (2005) and Carlsson and Martinsson (2006).

[6] We refer here to the second phase of the study conducted by Accent Marketing and Research on behalf of the UK regulator [Ofgem, 2004b].

#### 6.2.2.1 Contingent valuation analyses

Contingent valuation analyses comprise two basic steps: the survey itself and the analysis of the surveyed data. We illustrate them using the Italian customer survey as a practical example [Bertazzi et al., 2005].[7] In reality, a third important step is involved in customer surveys: the regulatory decision on the incentive rates. This decision depends, to a large degree, on the regulatory framework and on the regulator's judgement. For this reason, there is no guideline to be given, and the subject is only briefly mentioned below.

**The survey**

When conducting a contingent valuation analysis, a first decision to be made regards the target customer. Of course, this decision depends on the objective of the survey. In the Italian case, the objective was to define two different incentive rates to be employed in the reward and penalty scheme. Such incentive rates had to capture customer WTP for continuity of supply for two low voltage customer categories: domestic and non-domestic. Accordingly, the customer target was composed by all low voltage customers in the country.

The customer pool was then broken down into two major customer categories: domestic and non-domestic (note that the identification of different customer categories allows the regulator, if necessary, to develop different survey instruments for each category). Next, the sample for the first category was defined, stratifying the reference universe (22 million households) to reflect urban, suburban, and rural locations, as well as geographical differences (North West, North East, Centre, South and Island). The sample for the non-domestic category was created from the reference universe (3.8 million businesses) using company size as an additional variable. This was motivated by the large heterogeneity observed in the reference universe.

Sample size identification and sample selections require particular attention in order to secure representative and statistically meaningful responses for all customer categories. Note that, generally, fixed development and other costs of conducting a survey are high compared to the incremental costs per customer surveyed [Wacker and Billinton, 1989].

Another decision that has to be made at an early stage regards the practical form of the interview. Surveys can be conducted by mail, by telephone, or by means of face-to-face interviews. Since customer surveys on interruption costs and/or WTP tend to have considerable detail and require respondent reflection, telephone surveys are not generally well-suited for this purpose [Wacker and Billinton, 1989]. In the Italian case, although in other countries customer surveys had been successfully

[7] The practical reason for this choice is that the authors are familiar with this survey. Note that the Italian survey was implemented in collaboration with two Norwegian institutions. Their contribution was particularly significant in adapting the valuation model used in Norway to the Italian context.

conducted by mail, the expected response rate for a postal interview was deemed to be extremely low. Therefore, the interviews were conducted face-to-face.

The most important choice regards the valuation method for interruption costs. The most obvious approach is a direct solicitation of the customers' costs for given interruption characteristics. As already mentioned, this approach provides reliable results in those situations where costs tend to be tangible, directly identifiable and quantifiable. In other words, this approach is well suited for the industrial and business sectors [Wacker and Billinton, 1989]. As for domestic customers, other evaluations are preferable, such as WTP and WTA. It is possible (and often desirable) to include more than one valuation method. As for domestic customers, the Italian survey quantified direct costs, WTP and WTA.

A contingent valuation approach requires respondents to quantify costs (WTP and WTA) in a few different hypothetical interruption scenarios. Typically, a reference case is taken as a starting point, for example a two hour interruption starting at 18:00 on a winter evening. Variations in duration and time of occurrence (day of the week, season) are normally also used. Other variations in interruption characteristics can be adopted as well. For instance, the frequency may vary: a monthly interruption, an annual interruption, and so on. Notified and un-notified interruptions can also be considered.[8] Although it would be interesting to investigate all possible factors that might affect the cost of the interruption, the length of the questionnaire is of course limited by the degree of effort respondents are believed to be willing to accept, and by the complexities of the data analysis that will follow the survey [Wacker and Billinton, 1989]. The Italian questionnaire used a 2-hour interruption scenario and varied the duration (3 minutes, 1 hour, 4 hours and 8 hours), as well as the time of the week (weekdays and weekends) and the season.

As far as direct costs are concerned, respondents are simply requested to evaluate their direct costs. Guidance can be offered on what should and should not be included in the cost estimate so that the results are consistent across respondents.

As for WTP, customers are typically asked to imagine that a company offers a back-up electricity supply which can be purchased in case of an interruption of the main supply. The questionnaire then explores how much the respondents are willing to pay for the hypothetical back-up system in order to avoid the interruption. The estimates are made by the respondents for the reference case scenario as well as for the other selected scenarios. Sometimes, the questions can be formulated in terms of how much more or less (expressed as a percentage) they are willing to pay with respect to the reference case.

---

[8] Carlsson and Martinsson (2005) find that WTP among Swedish households is significantly higher for un-notified interruptions. Apart from obvious reasons (customers can prepare for a notified interruption and reschedule their activities), uncertainty about the duration of the interruption creates significant, additional inconvenience to for domestic customers. In other words, providing information about the expected length of an interruption would would be very important. Moreover, this also means that, when asked about their WTP to reduce an unplanned interruption for a given scenario of known length (such as in contingent valuation analyses), customers are actually underestimating their responses.

WTA is estimated as the monetary amount that respondents would consider satisfactory to receive, if the company supplying electricity offered a discount on their bill for each interruption event of the type considered in the questionnaire.

It is important to note that, in theory, WTP and WTA should be the same: the amount a rational individual is willing to pay for a replacement service should be no more than the damage one sustains when the service is deficient. Similarly, a discount equal to the loss suffered should be sufficient to provide full recovery for the damage. However, in practice, WTP figures are often significantly smaller than WTA figures.[9]

Finally, the questionnaire should obviously include questions regarding customer demographics, energy consumption, contractual capacity, and other similar data. In addition, a few general questions normally precede the economic valuation and allow respondents to qualitatively describe the severity and inconvenience of supply interruptions. It can be seen that developing the actual questionnaire generally requires considerable work. Moreover, small-scale testing with a small number of customers might be necessary. Questionnaires used by other researchers always need adjustments. Often, consultants with experience in survey techniques are hired and the field study is conducted by an external market research company.

### Analysis of the results

Surveyed data are analyzed using standard statistical techniques, including calculation of mean values and standard deviations, and correlations among variables. Raw data are subject to outlier analysis (anomalous values, exceeding specific thresholds calculated on the basis of the distribution of responses), and anomalous values are censorized (suppression of questionnaires featuring anomalous or missing values) and winsorized (repeated replacement of extreme values with other values, until these fall within an acceptable interval) on the basis of both statistical and logical consistency criteria.

Then, descriptive statistics are compiled. Typically, direct costs (WTP and WTA) are reported for various customer categories as a function of interruption duration, and they may be compared to the results of previous analyses. For example, direct costs for the different scenarios and customer categories in the Italian survey are represented in Figure 6.2.

As expected, direct costs increase with the duration of the interruption and with the contractual capacity (and annual consumption) of the customer. In particular, the interruption costs in Figure 6.2 refer to:

- the average Italian domestic customer (3 kW and 2800 kWh/year);
- three non-domestic customer categories ($\leq$10 kW, 10 kW to 30 kW, and >100 kW).

[9] Beenstock et al. (1998) offer an explanation for this difference in values. First, customers have a resistance to change, regardless of whether the service is improved or deteriorates (the 'status quo' effect). Second, customers always value service improvements by some fraction of their value of deterioration ('loss aversion' or 'asymmetry' effect).

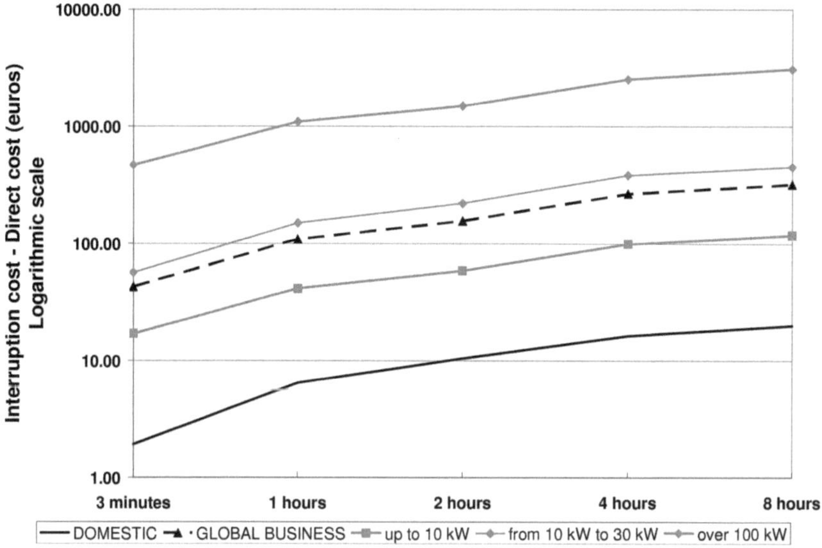


**Fig. 6.2.** Direct costs, Italian customer survey [Bertazzi et al., 2005]

Direct costs have a non-zero intercept (the cost of a three-minute interruption). This indicates that non-negligible costs are associated with the occurrence of the interruption (and not only with its duration). The curve then increases up to the maximum interruption duration of 8 hours (longer interruptions were not studied because they are unlikely to occur).

However, these values are of little use for the main objective of the survey. In the same way, they were not used in this form by electric companies. Allan and Billinton (2000) indicate that:

- for each respondent, a Customer Damage Function (CDF) was calculated, dividing the cost of the interruption (for the various interruption durations) by the annual energy consumed (or the peak demand);
- for each customer category, a Sector Customer Damage Function (SCDF) was calculated, weighting the above normalized costs according to annual energy consumed (or by peak demand). SCDF represent the cost an average consumer in the customer category would incur per kWh consumed annually. As raw data are obtained as a cost per interruption, SCDF are not the costs of the energy-not-supplied (this was often cause of misunderstanding);
- finally, knowing the customer mix for the area, the SCDF for the various customer categories were weighted according to their respective energy consumption within the area to give a cost function for the area. The series of values for the range of duration studied formed the Composite Customer Damage Function (CCDF) for the distribution area, which is, of course, system-dependent (unlike the SCDF).

By contrast, regulators are interested in the costs of the energy-not-supplied (i.e., of the energy that would have been consumed if the interruption had not occurred). To derive these values, regulators need to know load curves or, at least, to estimate them to a good degree of approximation. In addition, they need to arrive at an average cost of the energy-not-supplied per customer category (for different interruption durations).

For instance, in the Italian survey, the regulator estimates the energy consumption of each respondent $i$ in four different time-bands ($TE_{i,T}$): peak hours, high load, medium load, low load. The *cost of energy-not-supplied* ($C_{ENS,i,T}$) *for a respondent i in time-band T, for an interruption of duration r*, occurring in time-band T (of duration $h_T$), is thus given by:

$$C_{ENS,i,T}(r) = \frac{c_{i,T}(r)}{r \cdot \frac{TE_{i,T}}{h_T}}$$

in [euros/kWh], where $c_{i,T}(r)$ is the direct cost estimated with the survey for an interruption scenario of duration $r$, occurring in time-band $T$.

An *average cost of energy-not-supplied for respondent i, for an interruption of duration r*, is thus estimated weighting the above per time-band values, using appropriate weights for each time-band. Finally, the *cost of the energy-not-supplied for the customer category* (domestic and non-domestic) *for an interruption of duration r* is calculated as an average of the per respondent values, for all respondents within the same category. Table 6.2 shows these normalized direct costs (i.e., the costs of the energy-not-supplied) for the Italian survey.

Note that normalized direct costs decrease with duration: in terms of both direct and less tangible consequences, the initial impact and the early period tend to have the greatest effect. It is also interesting to note that the per-hour normalized costs diminish as the interruption duration increases: the increase in direct costs observed in Figure 6.2 is less than proportional to the energy-not-supplied.

Normalized WTP and WTA values are obtained in the same manner. These normalized values constitute the final output of the analysis of the results.

A final remark regards the comparison of these data across countries. While direct costs are often found to be of the same order of magnitude, normalized values are strongly influenced by the average energy consumption. *Ceteris paribus*, higher consumption patterns lead to lower normalized values and vice versa. Hence,

**Table 6.2.** Normalized direct costs, Italy

| | Domestic [Euro/kWh] | Business [Euro/kWh] |
|---|---|---|
| 1 hour | 25.34 | 117.98 |
| 2 hours | 20.41 | 83.79 |
| 4 hours | 15.73 | 67.18 |
| 8 hours | 9.68 | 40.01 |

when comparing costs of energy-not-supplied, it is important to state the average consumption levels assumed in the analysis (see also Section 4.4.5).

### The regulatory decision

After receiving the results of the analysis, the regulator still needs to transform this information into incentive rates for the reward and penalty schemes. This task requires considerable effort and, in part, also calls for discretionary decision-making. There is really no established procedure to follow.

In Norway and Italy the structure of the incentive scheme is symmetric around the performance standard and, thus, the incentive rate had to be the same for both rewards and penalties. By contrast, an extensive spread was observed between WTP and WTA (WTA was 4 to 7 times higher than WTP). Given these constraints, regulators have chosen incentive rates within a close range of the average of the WTP and WTA values. In particular, they have focused on the WTP and WTA values that resulted from the most representative interruption scenario (i.e. the scenario with an interruption duration that coincided with the average SAIDI). Indeed, the large difference between WTP and WTA is one of the major problem with contingent valuation analyses.

#### 6.2.2.2 Conjoint analyses

According to Ofgem (2004b), a conjoint analysis (or stated preference analysis) is a method of obtaining relative importance and willingness to pay for changes in service. The basis for this technique is the concept of 'utility'. Each product or service conveys a benefit to the customer; in his or her purchase behaviour, the customer seeks to maximise that benefit and makes choices accordingly. However, the utility of a product or service is the sum of the utilities of the features or attributes of the product/service. Each of these features conveys a benefit to the purchaser (or a disadvantage, in the case of price). When making choices, customers trade off the bundle of benefits of one product or service against those of an alternative product or service. Conjoint analyses seek to measure the utility associated with a product or service and to derive the particular value associated with each of its attributes.

In a conjoint analysis, the measurement is undertaken in an experiment or exercise that constitutes the principal part of the interview. Respondents are offered a series of choices between two packages, each of which describes the service in terms of the attributes for which measurement is required. Respondents are asked to make a choice between one of the two packages in each pair. The way in which the packages are designed conforms to a statistical design and may therefore be analysed.

The output of the analysis is a utility model comprised of utility weights for the attributes being examined. The weights show the relative preferences for each attribute and the strength of those preferences. When cost is included, its utility weight can be used to derive a willingness to pay for other attributes.

**Table 6.3.** Example of a pair-wise choice [Ofgem, 2004b]

| Choice A | Choice B |
|---|---|
| Fewer interruptions in rural areas: from an average of 8 to 5 over 5 years | The same number of interruptions in rural areas: an average of 8 in 5 years |
| Fewer interruptions in urban areas: from an average of 3 to 2 over 5 years | Fewer interruptions in urban areas: from an average of 3 to 2 over 5 years |
| No change in average interruption duration: an average of 114 minutes | Longer average interruption duration: from an average of 114 to 134 minutes |

The main difference as compared to the surveys described above is in the core section of the questionnaire, or the 'stated preference exercises'. In the UK case, the survey was designed to obtain customers' priorities between, and willingness to pay for, a list of potential changes in services. Note that a previous survey had already provided indications on customer priorities and expectations. The stated preference analysis could thus focus on a limited number of potential changes.

The potential changes in service included (but were not limited to):

- an increase (or decrease) in the number of interruptions in rural areas over 5 years: up to 3 more (or fewer) interruptions than the distribution company's current average;
- an increase (or decrease) in the number of interruptions in urban areas over 5 years: up to 3 more (or fewer) interruptions than the distribution company's current average;
- longer (or shorter) average interruption durations: up to 20 minutes longer (or shorter) than the distribution company's current average.

All variables were arranged in four stated preference exercises. Each exercise involved a series of six to eight pair-wise choices. An example of a pair-wise choice is given in Table 6.3: the respondent chooses his or her preferred package of improvements from the two options proposed.

Three variables were included in each of the first three exercises, and the final exercise traded bundles of improvements for changes in the size of the bill. This provided the monetary values and overall priorities between all the changes tested.

As for the survey described earlier, surveyed data are then analyzed and the results of the analysis are used in the regulatory decision. Note that, with this type of survey, information is obtained regarding both customer WTP for higher or lower levels of interruption frequency and duration, and for customer expectations in terms of performance standards.

## 6.3 Audits on data reported by companies

The robustness of service quality regulation primarily depends on the consistency and accuracy of information that regulated companies are reporting. Throughout this handbook we have described how regulators can introduce instructions and guidance on measuring and reporting of data. We have also mentioned that regulators may introduce periodical audits of measurement systems and reported data. Finally, we have explained the role of SCADA systems in continuity of supply data validation (Section 4.1.2.4). Here we focus on the auditing procedure, describing its key aspects and providing two practical examples.

Audits have two key objectives: (*i*) to verify that regulated companies are correctly applying the instructions and guidance for measuring and reporting service quality data, and (*ii*) to verify that companies meet specific minimum levels of accuracy while performing these tasks. Such minimum levels of accuracy are defined by the regulator, together with actions to be taken in case of inaccuracy.

When designing an audit procedure, regulators define three fundamental elements:

- instructions for the companies to ensure the traceability of all the reported data;
- indicators of accuracy and minimum acceptable levels of these indicators;
- corrective actions to be taken in case of non-compliance with the minimum levels (and possibly, associated financial penalties).

In addition, regulators define the process, the timetable, and the focus of the auditing procedure.

Regulators normally require that internal and/or external audits be carried out with a given periodicity (often once per year). Internal audits are conducted by trained staff of the regulated companies who have no role in performing any of the measuring and reporting functions. External audits can be conducted by a designated department of the regulatory authority or by independent consultants on behalf of the regulator (to whom they must report). Both commercial quality and continuity of supply data may be subject to audits. The details of an auditing procedure will strongly depend on the regulatory instructions regarding measurement and reporting. Note that the auditing procedure is normally the result of the joint effort of all interested parties, and it evolves in time as more experience is acquired.

Indeed, the overall design of an auditing procedure can be rather complex. Regulators have to address a large number of different issues, which range from maintenance of registers to corrective actions to be implemented in case of inaccuracies. For the purpose of this handbook, we focus on continuity of supply data and we focus only on the main aspects of the procedure: the indicators of accuracy, the minimum levels of accuracy, and corrective actions. To illustrate these points, we rely on two practical examples, concerning Italy and the UK. For information on all the other aspects of the auditing procedures, we refer the reader to the actual decision documents [AEEG, 2004, Ofgem, 2005b]. It is advisable to monitor the evolution of the described procedures: revisions and updates may contain significant insights on a number of practical issues.

### 6.3.1 Italian auditing procedure

In Italy, audits are carried out by the staff of the regulatory authority and normally last one or two days. Each year the audits concern a randomly selected sample of districts, and they are conducted on site at the corresponding SCADA operating centre (each operating centre controls more than one district). A number of interruption events are both randomly sampled and strategically selected for the audit, mainly among those registered by the SCADA system. As for interruptions originating on the MV and HV portions of the network, the audit is conducted by comparing the register with the information automatically registered by the SCADA system. In general, 10% of the total number of HV and MV events are verified during an audit. As for interruptions originating on the LV portions of the network, the company is required to keep a register for supply interruption-reporting calls. Thus, the audit can also concern those events; however, their contribution to the district SAIDI is generally quite small (around 10% of the average interruption duration over the year), and they have been given less attention (at least in the past).

For the sampled interruptions, the auditors verify that:

- the interruptions were actually reported in the register and that the information regarding the interruptions were registered according to the regulatory instructions (registration accuracy);
- the statistical indicator (SAIDI) computed by the auditors is equal to the statistical indicator computed by the company (precision);
- the regulatory causes of exemptions are indicated correctly and are documented according to the regulator's instructions (correctness).

Accordingly, the Italian regulator defines three evaluation indices [AEEG, 2004]:

- *Registration accuracy index*: a conventional scale has been defined to measure accuracy in recording interruptions, a scale that attributes different weights to different types of inaccuracies. The worst type of inaccuracy occurs when the company internal procedures for recording long interruptions are not robust and interruption events can go unrecorded. Other, less severe types of inaccuracy occur, for example, when the procedures for recording the number of affected consumers or the duration of the interruption are not sound. The index assumes values in the 0-1 range, where 1 indicates that the recording procedures are accurate.
- *Precision index*: this index assumes a positive value when the SAIDI, computed by the company on the sampled interruptions is lower than the SAIDI computed by the auditors on the same interruptions, and vice versa. The precision index is equal to 0% when the company's calculation precisely matches that of the auditors.
- *Correctness index*: exemptions due to *force majeure*, third party actions, and events originating on the transmission network and on the HV voltage portions of the distribution network need to be appropriately documented, according to the regulatory instructions. The correctness index assumes values in the 0-100%

range. A zero value indicates a completely incorrect attribution of the regulatory causes for an exemption. A 100% value indicates absolute correctness.[10]

The data provided by the company are considered 'valid' when: (*i*) the precision index is less then or equal to 3% (a maximum approximation of ± 3% is allowed) and (*ii*) the correctness index assumes values indicating that incorrectly excluded interruptions contribute to less than the 3% of the average interruption duration (SAIDI) as calculated by the company.

When one of the two above limits is not respected, the 'regulated indicator' reported by the company (SAIDI for the district, net of exemptions) is replaced by a 'presumed' SAIDI, calculated by the auditors on the basis of the information collected during the audit. The 'presumed' SAIDI is then used to calculate financial penalties, if they apply. When financial rewards are due, according the 'presumed' SAIDI, only 50% of the total amount is actually charged to the company.

In addition, when the accuracy index is less than or equal to 0.95, financial incentives are reduced (or penalties are increased) for all districts controlled by the audited SCADA centre. In the case of incentives, the reduction is calculated by multiplying the incentives by the accuracy index (with a cap at one half of the initial value). Similarly, in the case of penalties, the increase is calculated by dividing the penalties by the accuracy index (with a cap at twice their initial value).

Finally, administrative financial sanctions are provided for and may be applied in the event companies provide false data.

The final report concerning the audit can be consulted by the interested company upon request.

### 6.3.2 UK auditing procedure

The UK regulator normally appoints consultants to work jointly with its own staff to carry out audits of electricity distribution network operators' (DNO) interruption reporting. The audits are conducted annually for all 14 DNO and, in general, they last approximately 1 week. Audits concern the systems that each DNO has in place to measure incidents, statistical indicators on the number and duration of the interruptions, and the accuracy of the information reported by each DNO on these matters [Ofgem, 2006a].

Before going into the details of the procedure, it is necessary to clarify a few terms [Ofgem, 2005b].

- *Incidents* are defined as any occurrence on the DNO's distribution system or other connected distributed generation, transmission or distribution system which results in an interruption of supply to customer(s) for three minutes or longer.
- The *statistical indicators* used in the UK, respectively for the number and duration of long interruptions, are the following:

[10] This index is, by default, equal to 100% if the company has opted for the statistical methodology. See Section 4.6.

- *customer interruptions* (CI): the number of customers whose supplies have been interrupted per 100 customers per year over all incidents, where an interruption of supply lasts for three minutes or longer, calculated as:

$$\frac{\text{sum of the number of customers interrupted for all incidents} \cdot 100}{\text{total number of customers}}$$

- *customer minutes lost* (CML): calculated as in the SAIDI indicator.

- In the UK, the regulator defines low voltage as being less than 1 kV, while high voltage is between 1 kV and 22 kV, and extra high voltage is between 22 kV and 132 kV.

The auditing procedure is structured as 3-stage process (see Figure 6.3):

- Stage 1: audit of the measurement systems;
- Stage 2: audit of incident reporting and combined accuracy;
- Stage 3: additional audit of incident reporting and combined accuracy.

### Stage 1: Audit of measurement systems

In Stage 1, the accuracy of DNO measurement systems is assessed by [Ofgem, 2006a]:

- looking at several qualitative aspects: the way in which DNO have counted customers in their connectivity models and the underlying assumptions that they have used to link customer information to their network models, or whether DNO have correctly applied the definitions given in the regulatory instructions and guidance; and

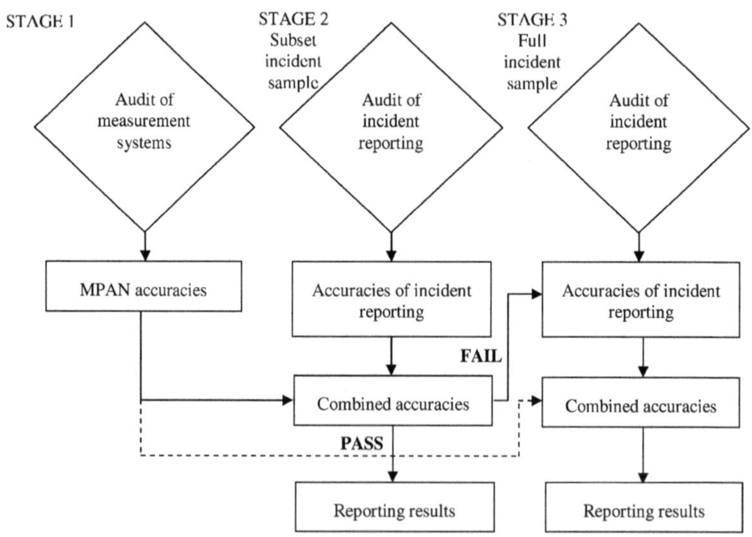


**Fig. 6.3.** Audit process in the UK [Ofgem, 2006a]

- calculating a so-called 'MPAN accuracy' (DNO identify customers from Metering Point Administration Numbers, such that one customer is identified at each connection point).

In order to estimate the overall MPAN accuracy, the *HV and above MPAN accuracy* and the *LV MPAN accuracy* are calculated using the following formulae [Ofgem, 2005b]:

HV and above MPAN accuracy =

$$= \left(\frac{\text{total number of MPAN assigned to true feeders at HV and above}}{\text{total number of MPAN}}\right) \cdot 100$$

LV MPAN accuracy =

$$= \left(\frac{\text{total number of MPAN assigned to true feeders at LV}}{\text{total number of MPAN}}\right) \cdot 100$$

In a DNO connectivity model, LV MPAN attached to true feeders (as opposed to dummy feeders or postcodes) will be recorded as losing supply when the feeder or substation to which they are attached becomes disconnected. Therefore, they contribute to CI and CML. By contrast, LV MPAN not connected to true feeders do not contribute to CI and CML. At the higher voltage levels (HV and above), MPAN attached to dummy LV feeders will lose supply when the substation becomes disconnected and, in this case, contribute to CI and CML [Ofgem, 2006a].

The *Overall MPAN accuracy* is then calculated as [Ofgem, 2005b]:

$$\text{Overall MPAN accuracy} = (\text{LV MPAN accuracy} \cdot \text{average \% LV contribution to CI} + \text{HV and above MPAN accuracy} \cdot \text{average \% higher voltage contribution to CI}) \cdot 100$$

### Stage 2: Audit of incident reporting and combined accuracy

At the end of the reporting year, after the interruption data have been submitted, and before the audit takes place, the regulator selects a sample of 150 incidents, split between HV and above and LV, according to their respective contribution to CI and CML (with a minimum of 50 LV incidents).

An 'Overall data set' of 50 HV and above incidents and 30 LV incidents is selected. The 30 LV incidents are identified as the 'LV data set'. A detailed examination of each of the incidents in the overall sample follows. The examination includes an assessment of [Ofgem, 2006a]:

- how the number of customers affected by each restoration stage of each incident as reported by the DNO relates to both the audit trail (information generated at the time of the incident recorded in field records, switching logs or other measurement systems) and the number of customers shown on the DNO's connectivity model;
- how the reported duration of each stage of each incident compares with the audit trail for the incident that occurred (e.g., the time of the first customer call registered in the call logs and restoration times recorded in field records, switching logs or other measurement systems);

- whether each incident is captured by the measurement systems by comparing customer and incident reports, and whether logged network events relate to relevant incident reports; and
- comparing the location of each incident within the distribution networks with the representation in the measurement systems.

'Audited' values for CI and CML are thus calculated for each incident, (removing outlying results). These values are compared against the original values reported by the DNO to measure the level of accuracy/inaccuracy. In particular, the accuracies of incident reporting for CI (and CML) for both the Overall and the LV datasets are calculated according to the following formulae [Ofgem, 2005b]:

Overall accuracy of incident reporting CI (CML) =

$$= \frac{\text{sum of reported CI (CML) in overall dataset}}{\text{sum of audited CI (CML) in overall dataset}}$$

LV accuracy of incident reporting CI (CML) =

$$= \frac{\text{sum of reported CI (CML) in LV dataset}}{\text{sum of audited CI (CML) in LV dataset}}$$

Finally, the *combined accuracies of reporting* are calculated using the following formulae [Ofgem, 2005b]:

Combined Overall CI (CML) accuracy of reporting = 100 - abs [100 - MPAN Overall accuracy · Overall accuracy of incident reporting CI (CML) · 100]

Combined LV CI (CML) accuracy of reporting = 100 - abs [100 - LV MPAN accuracy · LV accuracy of incident reporting CI (CML) · 100]

If the combined Overall and LV accuracy results meet the threshold levels (respectively, 97% and 92%), the DNO is deemed to have met the minimum accuracy and the audit is complete (See Table 6.4).

If either the combined Overall CI or CML accuracies are below the threshold levels, then the audit proceeds to Stage 3. Similarly, if either the combined LV CI or CML accuracies are below the threshold levels, then the audit proceeds to Stage 3.

**Table 6.4.** Minimum levels of accuracy, UK [Ofgem, 2006a]

| | Overall | | LV | |
|---|---|---|---|---|
| | Stage 2 - Subset incident sample | Stage 3 - Full incident sample | Stage 2 - Subset incident sample | Stage 3 - Full incident sample |
| Customer interruptions (CI) | 97% | 95% | 92% | 90% |
| Customer minutes lost (CML) | 97% | 95% | 92% | 90% |

### Stage 3 Additional audit of incident reporting and combined accuracy

In Stage 3, the Overall dataset consists of all 150 HV and above and LV incidents. Similarly, the LV dataset consists of all LV incidents (minimum 50).

The accuracies of incident reporting and the combined accuracies are calculated using the same method described in Stage 2.

The threshold level of combined Overall accuracy (CI and CML) is set at 95%. The threshold level of combined LV accuracy (CI and CML) is set at 90% (See Table 6.4).

If the DNO fails to meet these minimum levels of accuracy required for the reporting of CI and CML, the regulator will make the appropriate adjustments to performance (the data will be made 100% accurate) [Ofgem, 2005b].

At the end of the process, the auditors prepare a final report, with the calculation details and accuracy results for each DNO. These reports are made available on the regulatory authority website.

# Bibliography


[AEEG, 2004] AEEG, 2004. Testo integrato delle disposizioni dell'Autorità per l'energia elettrica e il gas in materia di qualità dei servizi di distribuzione, misura e vendita dell'energia elettrica per il periodo di regolazione 2004-2007. Decision 4/04. Available (in Italian) from: [http://www.autorita.energia.it].

[Ajodhia and Hakvoort, 2005] Ajodhia, V., Hakvoort, R., 2005. Economic regulation of quality in electricity distribution networks. Utilities Policy 13 (3), 211–221.

[Allan and Billinton, 2000] Allan, R., Billinton, R., 2000. Probabilistic assessment of power systems. Proceedings of the IEEE, 88 (2), 140–162.

[Baldwin and Cave, 1999] Baldwin, R., Cave, M., 1999. Understanding regulation. Theory, strategy and practice. Oxford University Press, Oxford.

[Beenstock et al., 1998] Beenstock, M., Goldin, E., Haitovsky, Y., 1998. Response bias in a conjoint analysis of power outages. Energy Economics, 20 (2), 135-156.

[Bertazzi et al., 2005] Bertazzi, A., Fumagalli, E., Lo Schiavo, L., 2005. The use of customer outage cost surveys in policy decision-making: the Italian experience in regulating quality of electricity supply. In: CIRED 2005, International conference on electricity distribution, Turin, Italy.

[Bollen, 2000] Bollen, M., 2000. Understanding power quality problems. Voltage sags and interruptions, IEEE Press, Piscataway, NJ, USA.

[Bollen and Verde, 2007] Bollen, M., Verde, P., 2007. A framework for regulation of rms voltage and short-duration under and overvoltages. Mimeo.

[Brekke, 2006] Brekke, K., 2006. The Norwegian regulation on quality of supply. The Norwegian Water Resources and Energy Directorate. Written communication to the authors.

[Carlsson and Martinsson, 2005] Carlsson, F., Martinsson, P., 2005. Willingness to pay among Swedish households to avoid power outages. Elforsk rapport 05:04. Available from: [http://www.elforsk.se].

[Carlsson and Martinsson, 2006] Carlsson, F., Martinsson, P., 2006. How much is too much? An investigation of the effect of the number of choice sets, starting point and the choice of bid vectors in choice experiments. Elforsk rapport 06:21. Available from: [http://www.elforsk.se].

[CENELEC, 1999] CENELEC, 1999. Voltage characteristics of electricity supplied by public distribution systems. European Standard EN 50160: 1999–1-1.

[CER, 2006] CER, 2006. Quality of service incentive mechanisms during the 2006-2010 price control period for ESB PES & ESB DSO. Response and decision paper CER/06/107. Available from: [http://www.cer.ie].

[CEER, 2003] CEER, 2003. Second benchmarking report on quality of electricity supply. Available from: [http://www.ceer-eu.org].

[CEER, 2005] CEER, 2005. Third benchmarking report on quality of electricity supply. Available from: [http://www.ceer-eu.org].

[CNE, 2000] CNE, 2000. Regulating transmission, distribution, trading and supply activities and authorization procedures for electric power installations. Royal Decree 1955/2000. Unofficial English translation. Available from: [http://www.cne.es].

[DTe, 2002] DTe, 2002. Yardstick competition. Regional electricity network companies, second regulatory period. Information and consultation document. Unofficial English translation. Available from: [http://www.dte.nl].

[ERD, 2000] ERD, 2000. Contrat d'accès au rèseau public de distribution d'èlectricitè pour un site consommateur èligible raccordè en moyenne tension. Available (in French) from: [http://www.edfdistribution.fr].

[ERGEG, 2006] ERGEG, 2006. Towards voltage quality regulation in Europe. ERGEG public consultation paper E06-EQS-09-03. Available from: [http://www.ergeg.org].

[EC, 1985] European Commission, 1985. Liability for defective products. Directive 85/374/EEC. Available from: [http://www.europa.eu].

[Fumagalli et al., 2004] Fumagalli, E., Black, J., Vogelsang, I., Ilic, M., 2004. Quality of service provision in electric power distribution systems through reliability insurance. IEEE Transactions on Power Systems, 19 (3), 1286–1293.

[Fumagalli et al., 2006] Fumagalli, E., Lo Schiavo, L., Salvati, S., Secchi, P., 2006. Statistical identification of major event days: an application to continuity of supply regulation in Italy. IEEE Transactions on Power Delivery, 21 (2), 761-767.

[Gammelgard and Larsson, 2003] Gammelgard, M., Larsson, M., 2003. The challenge of regulating natural monopolies in electrical distribution - experiences from Sweden. In: CIRED 2003,

International conference on electricity distribution, Barcelona, Spain.

[HEO, 2004] HEO, 2004. Guaranteed Standards in Hungary. Extract from HEO resolution - 2004. Minimum quality requirements for customers on the area of the Licensee (Distribution and Public Supplier). Unofficial English translation. Written communication to the authors.

[HEO, 2006] HEO, 2006. Overall Standards in Hungary. Extract from HEO Resolutions year 2003 and 2006. Unofficial English translation. Written communication to the authors.

[IEC, 2003] IEC, 2003. Electromagnetic compatibility (EMC) - Part 4-30: Testing and measurement techniques - Power quality measurement methods. IEC 61000-4-30. IEC, Geneva, Switzerland.

[IEC, 2005] IEC, 2005. Electromagnetic compatibility (EMC) - Part 3-3: Limits - Limitation of voltage changes, voltage fluctuations and flicker in public low-voltage supply systems, for equipment with rated current <= 16 A per phase and not subject to conditional connection. IEC 61000-3-3. IEC, Geneva, Switzerland.

[IEEE, 2004] IEEE, 2004. IEEE Guide for Electric Power Distribution Reliability Indices. IEEE Standard 1366-2003. IEEE, New York, NY.

[ITIC, 2000] ITIC, 2000. ITI (CBEMA) Curve. Information Technology Industry Council. Available from: [http://www.itic.org].

[Langset et al., 2001] Langset, T., Trengereid, F., Samdal, K., Heggset, J., 2001. Quality adjusted revenue caps: a model for quality of supply regulation. In: CIRED 2001, International conference on electricity distribution, Amsterdam, The Netherlands.

[Lo Schiavo et al., 2005] Lo Schiavo, L., Malaman, R., Villa, F., 2005. Continuity of electricity supply regulation driven by economic incentives: does it work? The Italian experience. In: CIRED 2005, International conference on electricity distribution, Turin, Italy.

[Nilsson, 2005] Nilsson, P., 2005. Quality control in the Swedish regulation and balance between network charges and quality. In: CIRED 2005, International conference on electricity distribution, Turin, Italy.

[NRS, 2004] NRS, 2004. Electricity Supply - Quality of supply - Part 2: Voltage characteristics, compatibility levels, limits, and assessment methods. NRS 048-2:2003. NRS, Pretoria, South Africa.

[NVE, 2005] NVE, 2005. The Norwegian directive on quality of supply. The Norwegian Water Resources and Energy Directorate. ISSN: 1501-2840. Available (in Norwegian) from: [http://www.nve.no].

[Ofgem, 2001a] Ofgem, 2001a. Guaranteed and overall standards of performance. Final proposals. Decision 09/01. Available from: [http://www.ofgem.gov.uk].

[Ofgem, 2001b] Ofgem, 2001b. Information and incentives project. Incentive schemes: Final proposals. Decision 78/01. Available from: [http://www.ofgem.gov.uk].

[Ofgem, 2000a] Ofgem, 2002a. Information and incentive project. Report on medium term network performance monitoring. Available from: [http://www.ofgem.gov.uk].

[Ofgem, 2002b] Ofgem, 2002b. Ofgem consultation policy. Available from: [http://www.ofgem.gov.uk].

[Ofgem, 2003] Ofgem, 2003. Expectations of electricity DNOs and WTP for improvements in service - Stage 1 Quantitative research findings. Consultation document 110/03. Available from: [http://www.ofgem.gov.uk].

[Ofgem, 2004a] Ofgem, 2004a. Electricity distribution price control review - Final proposal. Decision 265/04. Available from: [http://www.ofgem.gov.uk].

[Ofgem, 2004b] Ofgem, 2004b. Electricity distribution price control review - Appendix - Consumer expectations of DNOs and WTP for improvements in service report (prepared for Ofgem). Consultation document 145f/04. Available from: [http://www.ofgem.gov.uk].

[Ofgem, 2005a] Ofgem, 2005a. The Electricity (Standards of Performance) Regulations 2005. Statutory Instruments 2005 No. 1019. Available from: [http://www.opsi.gov.uk].

[Ofgem, 2005b] Ofgem, 2005b. Quality of service regulatory instructions and guidance version 5. Decision 94/05. Available from: [http://www.ofgem.gov.uk].

[Ofgem, 2005c] Ofgem, 2005c. Guidance on impact assessments - revised guidance. Document 148/05. Available from: [http://www.ofgem.gov.uk].

[Ofgem, 2006a] Ofgem, 2006a. Quality of service incentive scheme audit of interruptions reporting 2005/06 - Final Report. Available from: [http://www.ofgem.gov.uk]

[Ofgem, 2006b] Ofgem, 2006b. Guaranteed Standards: Ofgem guidance and proposals on best practice - Electricity distribution. Available from: [http://www.ofgem.gov.uk].

[Ofgem, 2006c] Ofgem, 2006c. Ofgem's decision on the future of the gas and electricity metering price controls. Decision 187/06. Available from: [http://www.ofgem.gov.uk].

[Rekettye and Orosdy, 2003] Rekettye, G., Orosdy, B., 2003 (January). Findings of the survey performed in 2002 on customer satisfaction with the electricity supply. Working paper. Department of marketing. Faculty of business and economics. University of Pècs, Hungary.

[Rivier and Gomez, 2003] Rivier, J., Gómez, T., 2003. A critical analysis of Spanish power quality regulation design. In: Market Design Conference, Stockholm, Sweden, June 16-17.

[Salvati et al., 2006] Salvati, S., Fumagalli, E., Lo Schiavo, L., Secchi, P., 2006. Analysis of the methodology adopted by the Italian regulatory authority for identifying major event days. In: Probabilistic Methods Applied to Power Systems, PMAPS, International conference, Stokholm, Sweden.

[Samdal et al., 2003] Samdal, K., Kjølle, G., Singh, B., Trengereid, F., 2003. Customers' interruption costs: what's the problem? In: CIRED 2003, International conference on electricity distribution, Barcelona, Spain.

[Sappington, 2005] Sappington, D., 2005. Regulating service quality: a survey. Journal of regulatory economics, 27 (2), 123-154.

[Seljeseth et al., 2005] Seljeseth, H., Sand, K., Samdal, K., 2005. Quality of supply regulation in Norway: going beyond EN 50160. In: CIRED 2005, International conference on electricity distribution, Turin, Italy.

[Tersztyanszky, 2003] Tersztyanszky, T., 2003. Methods and procedures requirements for monitoring and improvement of supply quality in Hungary. In: CIGRE' - IEEE Symposium on Quality and security of electric power delivery system, Montreal, Canada.

[Tersztyanszky, 2005] Tersztyanszky, T., 2005. First results of performance based regulation of supply quality in Hungary. In: CIRED 2005, International conference on electricity distribution, Turin, Italy.

[Wacker and Billinton, 1989] Wacker, G., Billinton, R., 1989. Customer cost of electric service interruptions. Proceedings of the IEEE, 77 (6), 919-930.

[Weisman, 2005] Weisman, D., 2005. Price regulation and quality. Information Economics and Policy, 17 (2), 165-174.

[Williamson, 2001] Williamson, B., 2001. Incentives for service quality: getting the framework right. The Electricity Journal, 14 (5), 62-7.

# Index

MIX
Papier aus verantwortungsvollen Quellen
Paper from responsible sources
FSC® C105338

If you have any concerns about our products,
you can contact us on
**ProductSafety@springernature.com**

In case Publisher is established outside the EU,
the EU authorized representative is:
**Springer Nature Customer Service Center GmbH**
**Europaplatz 3, 69115 Heidelberg, Germany**

Printed by Libri Plureos GmbH
in Hamburg, Germany